Managing Legal and Security Risks in Computing and Communications

PAUL D. SHAW

Butterworth-Heinemann
Boston Oxford Johannesburg Melbourne New Delhi Singapore

Library of Congress Cataloging-in-Publication Data
Shaw, Paul (Paul D.)
    Managing legal and security risks in computing and communications
/ Paul Shaw.
        p. cm.
    Includes bibliographical references and index.
    ISBN 0-7506-9938-8 (alk. paper)
    1. Computers—Law and legislation—United States. 2. Computer
networks—Law and legislation—United States. 3. Computer security—
United States. 4. Electronic data interchange—Law and
legislation—United States. 5. Data protection—Law and
legislation—United States. 6. Intellectual property—United
States. I. Title.
KF390.5.C6S53 1998
343.7309'99—dc21   97-35913
                                                                         CIP

British Library Cataloguing-in-Publication Data
A catalogue record for this book is available from the British Library.

The publisher offers special discounts on bulk orders of this book.
For information, please contact:
Manager of Special Sales
Butterworth-Heinemann
225 Wildwood Avenue
Woburn, MA 01801-2041
Tel: 781-904-2500
Fax: 781-904-2620

For information on all Butterworth-Heinemann publications,
contact our World Wide Web home page at: http://www.bh.com

10 9 8 7 6 5 4 3 2 1

Printed in the United States of America

Contents

v

Preface

The protection of computer and telecommunications technologies, computer data, and information has spawned a unique set of legal risks. Liability and litigation can arise from misuse and abuse of computer data bases, bulletin boards, e-mail and Web pages, electronic funds transfer systems, proprietary computer programs and from absent or ineffective disaster recovery procedures and data archiving. For a negligent act, a breach of privacy, or an infringement of intellectual property rights, tort, statute, and regulatory liabilities can attach to the organization, its directors and officers, management, employees, and agents. Protecting computers and information has become both a necessary business practice and a legal requirement.

Managing Legal and Security Risks in Computing and Communications is an up-to-date, plain-English guide to computer-related crime and protection/litigation risks. It examines information systems legal liability risks, including

- Failure to provide effective controls and security to prevent wrongful access to computers or networks;
- Using on-line systems in ways that violate another person's rights, including harassment, defamation, libel, or privacy;
- Failure to have a disaster recovery program;
- Failure to keep adequate records of and reasonable security for electronic transactions systems;
- Failure to make timely data backups and provide safe records storage;

- Improper use of vendor software or systems; and,
- Insufficient protection of proprietary information and intellectual property.

Managing Legal and Security Risks in Computing and Communications is written for information systems and computer security managers, chief information officers, legal counsel, compliance officers, risk managers, auditors, and others responsible for information systems protection and privacy. This book will help readers

- Quickly locate laws covering computer crime, intellectual property, and other statutes relevant to prosecution;
- Provide guidance on the law so management can be alerted about its possible liabilities and litigation risks concerning computer crime and protection of information assets;
- Give the legal facts and background essential to formulating computer protection policy;
- Foster a continuing dialogue with legal counsel on matters affecting information systems security and legal liabilities;
- Provide a legal reference for developing training materials that will help sensitize employees to legal compliance requirements and the importance of computer protection.

The first 12 chapters of the book cover law that requires organizations to have computer security. Safeguarding the confidentiality and privacy of information and the integrity and reliability of information systems is mandated in contracts, regulations, statutes, or in common law requirements for a standard of care or a fiduciary duty to protect a third party from harm or economic loss. Included are discussions of negligence, tort and contract law, liability of organizations, directors, officers, managers, employees, and agents.

Also covered are the U.S. sentencing commission guidelines on how to develop effective compliance programs that have ethics policies, codes of conduct, internal controls, and communications designed to make employees aware of their compliance obligations; plus, how to monitor, report and document compliance efforts.

Chapters focus on intellectual property, privacy, records management, and electronic commerce detail compliance mandates in the Foreign Corrupt Practices Act, the Electronic Funds Transfer Act, the Uniform Commercial Code, and consumer-related statutes, plus regulations issued by the Securities and Exchange Commission, Comptroller of the Currency, Federal Home Loan Bank Board, and other agencies.

Chapters 13 through 17 examine the law on computer-related crime. This law includes federal and state statutes on computer and telecommunications crime, as well as statutes on false statements and claims, theft, fraud, embezzlement, transport of stolen goods, malicious destruction of property, conspiracy, and forfeiture. Each statute has an analysis of its provisions and leading cases prosecuted under the statute.

This law also offers an organization a way to take action against those that breach its information systems security—specific legal tools for deterrence, prosecution, and recovery for damages.

The Appendix discusses industry and government standards on internal controls, computer security, and auditing; the Glossary defines important computer, security, and legal terms; and the Bibliography lists works on computer and intellectual property law, compliance, internal controls, and protection.

Computer Protection and Legal Liability Risks

The Computer Security Institute's "1996 Computer Crime and Security Survey" was composed of questions submitted by the FBI. The survey was sent to companies and government agencies, and responses were received from 428 organizations. The survey reveals that, even among companies that have some computer security, it may be inadequate in terms of policy, administration, and employee awareness and training.

Some of the survey findings include

- Of those responding, 41 percent have experienced some form of computer system intrusion or unauthorized use of their system during the last 12 months. Over 50 percent of those intrusions were traced to employees. Dial-in and Internet connections were another source of intrusions.
- Altering data was the most frequent attack reported by medical (36.8 percent) and financial (21 percent) institutions.
- Over 50 percent of respondents cited U.S. corporate competitors as a likely source of attacks ranging from eavesdropping to system penetration and reported that information sought would be of use to competitors.

- Attacks in which intruders forge a return address to gain access centered on independent hackers and external information brokers as likely sources.

Regarding computer security,

- Over 50 percent of the respondents have no written policy on how to deal with network intrusions; 60 percent have no policy for preserving evidence for criminal or civil proceedings; and 70 percent have no "warning" banner stating that computing activities may be monitored.
- Over 20 percent of the respondents did not know if they had been attacked. Less than 17 percent said they would advise law enforcement if they had been attacked. Over 70 percent cited fear of negative publicity as the primary reason for not reporting.

Expanding Civil and Criminal Liability

The term *litigious society* has been applied to the United States, and some states have been called *tort heaven* by lawyers. These terms describe both an attitude and a fact of life and business. It was announced three years ago that a trial lawyer's association formed two groups, one concentrating on computer vendor liability and the other on inadequate security litigation.

Not only are civil liability risks growing, corporations today are faced with lessened standards of criminal liability and a trend to criminalize regulations that were once essentially civil. Federal legislation initially intended to fight criminal racketeering and narcotics-related crimes has given prosecutors broad powers to indict, compel plea bargaining, and force cooperation of individual or corporate defendants.

Prosecution used to hinge on the concept of intent, but this has often been replaced by evidence of "willful blindness," "recklessness," "failure to perceive (a risk)," or "collective knowledge" for corporate liability. Added could be the expensive prospect of collateral prosecution and litigation or being twice charged, tried, and possibly fined or sentenced for a single violation.

Although legal costs can be exorbitant, litigation also can damage a company's reputation; harm employee, customer, and investor relations; and possibly diminish credit lines.

This cha⁻ ⸲er will examine several of the key legal elements that define liabil̤y and, specifically, those risks associated with computer protection.

Foreseeability, Due Care, and Negligence

"The probability of injury by one to the legally protected interests of another is the basis for the law's creation of a duty to avoid such injury, and foresight of harm lies at the foundation of the duty to use care and therefore of negligence. The broad test of negligence is what a reasonably prudent person would foresee and would do in the light of this foresight under the circumstances" (*American Jurisprudence*, 2d, Sect. 135). In possible criminal situations, "an actor . . . must anticipate and guard against . . . criminal misconduct of others" (ibid., Sect. 164). Further, it does not matter if the consequences of the act are unforeseen: "the test is not what the wrongdoer believed would occur; it is whether he ought reasonably to have foreseen that the event in question, or some similar event would occur."

Even though our focus here is on liability risks arising from an absence of or inadequate computer security, we must be aware that legal issues initially developed in one area, have a way of migrating into totally different fields. Such is the nature of foreseeability and its relationship to negligence and computer security.

Foreseeability cases typically are brought by victims of violent crime that occur on business premises. It is in this area that case law on foreseeability has produced judicial decisions ranging from conservative to increasingly liberal and expansive.

Liability can also arise from awareness of the imminent probability of specific harm to another. Here are several scenarios:

- A woman shot at a bar sues the man convicted of the shooting and the bar owner. The owner is sued because it is contended he should have known the bar's patrons were often armed.
- A man who was rehired by a company after serving prison time for killing a coworker kills another coworker. The company is sued by the coworker's survivors.
- An automobile is left unattended with the keys in the ignition in a high crime area. The car is stolen and is involved in an accident. The owner is sued for the injuries resulting from the accident.

- A woman is mugged and injured in a robbery close to an automatic teller machine (ATM) from which she had just withdrawn money. She sues the ATM operator.

These scenarios, in different ways, raise the legal issue of foreseeability. This issue centers on the legal obligation of one party to protect another, second party against foreseeable, intentional wrongs done by a third party. The connection among parties usually is that of a business or a vendor to a customer or client; the wrongful act is committed by a mugger, thief, or the like. Normally, in the past, responsibility or duty of care existed between the business and the customer; the owner was rarely blamed for an act inflicted on a customer by a criminal even if the act occurred on or near the owner's property.

Foreseeability has been put into regulation in a 1990 California law (California Financial Code, Sections 1300–1370) that covers ATM safety. In its procedures to evaluate the safety and security of ATM customers, operators must consider the incidence of violent crime in the neighborhood in which the ATM is located.

ATM crime is documented, and there have been court cases. For example, the Florida Bankers Association surveyed its members and found that 303 criminal incidents occurred at bank ATMs between 1991 and 1992.

The courts increasingly find a connection between parties, laying negligence on the business owner for failing to be aware of risks to customers from acts of third parties. The key phrase here is "aware of risks." The owner should be aware of the risks to his or her customers inherent in the business operations and the surrounding environment.

We are dealing here with several slippery words and concepts. Judge Richard Posner called *foreseeability* a "maddeningly vague" legal term. The same can be said for terms such as *reasonable care* and *awareness of risks*.

If all this is so vague, why bother with it? Because the concept of foreseeability increasingly is used in cases where negligence rests on absent or inadequate security.

Risk Awareness

How does one foresee a risk? It is hard to foresee a very specific risk; what matters in law is not that one be clairvoyant, but that one be aware of those risks that might affect one's customers and that one takes appropriate precautions to alleviate those risks.

This, in short, is a definition of the *duty of care*. But what should be your level of "knowing"? (Or, what should a reasonable person have known?) And what are the appropriate set of precautions? (Or, how much security should you have?)

Courts often evaluate foreseeability in terms of prior similar incidents and the circumstances surrounding the incident. Well, this provides a partial answer. It implies that you should be aware of prior incidents, and for computer systems, at several levels: global, local, and site specific. Let us look at awareness levels:

> *Global*—An example would be a study on virus or hacker attacks. At the national level, there are crime surveys by the U.S. Census Bureau, the FBI, and the National Criminal Justice Reference Center.

> *Local*—Forecasting techniques have made accurate crime projections for city blocks, census tracts, and zip codes. One company, CPA Index, Inc., produces color-coded maps depicting crime vulnerability for any location in the United States.

> *Site specific*—Risk assessments and security evaluations can be gathered from in-house loss reporting, security surveys, vulnerability analyses, computer system penetration tests, and internal audits.

All of these can be used to determine the foreseeability of wrongful acts; you could argue that the degree of security should be directly proportionate.

Conclusions

There is less likelihood of charges of and recovery for security negligence when appropriate precautions are taken. Remember, appropriate security does not mean just some security (that is likely to be called inadequate), but a well-designed and tested system. This is your best defense against charges of negligence and attendant liability.

Foreseeability should be evaluated in terms of

1. Prior similar incidents, their frequency and recentness;
2. Whether certain acts were "enhanced" by poor security;
3. The "totality of the circumstances," that is, the circumstances surrounding the incident.

There are no hard and fast rules as to what is legally foreseeable or what is reasonable security—it is situational, and it must be examined case by case.

To give you a better idea of the relationship among foreseeability, security, and negligence, the relevant legal concepts are outlined in the sections that follow.

Negligence Defined

Negligence is related to duty of care and defined by four elements:

1. A legally recognized duty to act as a reasonable person under the circumstances;
2. A breach of the duty by failing to live up to the standard;
3. A reasonably close causal connection, known as *proximate cause,* which includes cause in fact;
4. Actual loss or damages. (Keeton, 1984).

Further, "foresight of harm lies at the foundation of the duty to use care and therefore of negligence. The broad test of negligence is what a reasonably prudent person would foresee and would do in the light of this foresight under the circumstances." (*American Jurisprudence,* 2d, Sect. 135).

Duty of Care

The duty of care requirement involves

1. Foreseeability of harm to the plaintiff;
2. Closeness of the connection between the defendant's conduct and the injury incurred;
3. Degree of injury received;
4. Moral blame attached to the defendant's conduct;
5. Policy of preventing future harm.

"Duty of care is not sacrosanct in itself, but only an expression of the sum total of those considerations of policy which leads the law to say that the plaintiff is entitled to protection." (Keeton, 1984).

Notice

Liability also can arise from awareness of the imminent probability of specific harm to another. This is legally referred to as notice, or having specific knowledge concerning the existence of a fact or condition. Notice may give rise to a duty to protect or at least to investigate the situation.

Liability for Inadequate Security

An organization can be held liable for inadequate security if there is evidence that

1. The crime was similar to a previous crime committed on its property.
2. The organization did not take all economically feasible steps to provide a reasonable level of security.

The requirement of prior similar conduct involving a criminal act need not be proven to establish foreseeability. In a California case, *Isaacs v. Huntington Memorial Hospital,* foreseeability could be determined "in light of all the circumstances and not by a rigid application of a mechanical 'prior similar' rule." This liberal reading by the court was followed by another California case that was more expansive, *Southland* v. *Superior Court.*

Southland Corp. is the parent company of 7-Eleven convenience stores. This case involved one of its stores located in an area where there was little serious criminal activity. A vacant lot next to the 7-Eleven store was used by customers to park their cars when the store parking lot was full. Southland did not own or manage the vacant lot but did have the right from the landowner to let its customers use the lot. Of course, the store owners and staff were aware that their customers used the lot.

The store and the vacant lot became a teenage hangout; fights often broke out and the police were called by employees of the store. In one instance, several people were injured when they were attacked by some of the juveniles.

The Superior Court ruled that a jury had to decide whether Southland actually controlled the parking lot and the attendant liability question. Also, the question of the foreseeability of the criminal assault had to be resolved. The court said that issue "is not to be measured by what is more probable than not, but includes whatever is likely enough in the setting of modern life that a reasonably thoughtful person would take account of it in guiding practical conduct. One may be held accountable for creating even the risk of a slight possibility of injury if a reasonably prudent person would not do so."

Standards of Liability

Recent court cases and legislation have used some new terms to define standards of liability and burdens of proof:

Strict liability—The act only; this requires no proof of intent to commit the act.

Vicarious liability—In general, corporations are vicariously liable for the actionable conduct of their employees performed in the scope of their employment; this traditional doctrine applies to aiding and abetting a crime or a conspiracy to commit a crime; it includes acts with the knowledge and intention of facilitating the commission of a crime.

Derivative liability—The acts and intent of corporate officers and agents are imputable to the corporate entity.

Responsible corporate officer doctrine—This applies to any corporate officer or employee "standing in responsible relation" to a forbidden act. Liability can arise if the officer could have prevented or corrected a violation and failed to do so. Strict liability, the act only, and no mental element is involved. This is a critical doctrine with significant implications: an officer has a positive duty to seek out and remedy violations when they occur and a duty to implement measures that will ensure that violations will not occur. The responsible corporate officer doctrine derives from the Food, Drug and Cosmetic Act (see *United States* v. *Park*) and the Clean Water Act. Although normally applied to services and products that affect the health and well-being of the public, the doctrine easily could cover mental well-being, such as privacy. The doctrine forces the corporate officer to define which risks such a person should know, because he or she is likely to be held to an affirmative duty of care concerning those risks.

Willful blindness or indifference—Here the defendant intentionally avoids knowing a situation or act will incriminate. Willfulness is a disregard for the governing statute and an indifference to its requirements.

Flagrant organizational indifference—This is the conscious avoidance by an organization to learning about and observing the requirements of a statute.

"Ostrich" instruction—This permits a jury to infer guilty knowledge from a combination of suspicion and indifference to the truth. (see *United States* v. *Giovannetti*)

Rogue employee—One who, for his or her own benefit, commits an illegal act or whose conduct violates company policy and procedure despite in-place efforts to prevent such an act.

Collective or aggregate knowledge—Here, if the knowledge possessed by several employees adds up to their willful knowledge or yields a guilty state of mind, which is required by the statute, the organization can be liable for violating the statute. In *United States* v. *Bank of New England*, the bank was charged with violating the Currency Transaction Reporting Act, which requires a report to the IRS of transactions of more than $10,000 or multiple or structured transactions that total more than $10,000. A willful failure to report is a criminal violation. The court instructed the jury that if one employee knew of the requirement, the bank also had knowledge. However, the court went further, instructing the jury to first look at the bank as an institution with the bank's knowledge the sum of what all the employees' know within the scope of their employment. "So, if Employee A knows one facet of the currency reporting requirement, B knows another facet of it, and C a third facet of it, the bank knows them all." The Bank of New England was found guilty and the court of appeals upheld the collective knowledge instruction and its relevance to corporate criminal liability.

Other Sources of Potential Liability

Statutes, administrative rulings, and common law cases have expanded executives' and corporations' responsibilities and liabilities. At the federal level, the key legal rules for financial reporting are found in the Securities and Exchange Act, with its definitions of such things as the corporate insiders rule (10b-5); negligence involving new securities issues; and fraudulent misstatements, misrepresentations, or omissions in securities transactions (see *SEC* v. *Texas Gulf Sulpher*).

Other potential legal liabilities await the unwary. For example, failure to develop and implement contingency planning for data processing operations can have negative legal consequences for a company or its directors and officers. Here, problems may arise from the Foreign Corrupt Practices Act (FCPA) and its requirements for record keeping and internal controls.

State statutes also regulate corporate behavior and can have requirements similar to federal statutes. In some instances, state statutes may be preempted by federal law. One comprehensive statute, and perhaps a model for other states, is California's Corporate Criminal Liability Act. This law has a number of consumer protection features and stiff individual and corporate sanctions and fines.

Collateral Consequences of Convictions for Organizations

In 1991, the White Collar Crime Committee of the Criminal Justice Section of the American Bar Association (ABA) issued its final report on "Collateral Consequences of Criminal Convictions of Organizations."

The ABA report looked at the problem of collateral prosecutions and sanctions from three perspectives:

1. By examining the policy and constitutional issues surrounding those collateral consequences and their proper role in the overall sanctioning process;

2. By surveying and reporting on the extent and severity of the collateral consequences; and

3. By conducting an empirical survey of participants in the criminal justice system to develop a deeper understanding of how collateral consequences now work and how they relate to the criminal justice system.

The report consists of the legal and policy analysis of collateral consequences, a survey of those consequences, and a survey of criminal justice participants who have had experiences with corporate prosecutions. It concluded that collateral consequences matter a great deal and recommended "that the presentence report for a convicted organization include a 'Collateral Sanctions Report' identifying the collateral consequences that have been, or are likely to be, imposed on the organization."

Organizations face not only liability from criminal penalties but also a broad array of administrative sanctions and civil damages. One study, "Corporate Crime and Punishment: An Update on Sentencing Practice in the Federal Courts, 1988–90," found that "from 1984 to 1990 convicted corporate defendants paid criminal fines totaling approximately $215 million, but were assessed collateral sanctions totaling more than four times that amount, or $986 million."

Criminal penalties for organizations under the latest sentencing guidelines will rise steeply; for example, the maximum criminal penalty for antitrust violations has been raised from $1 million to $10 million.

The ABA report points out four reasons for increased coordination between criminal and civil sanctions:

1. Such coordination is consistent with a sound general theory of criminal law.

2. Recent court opinions, particularly the Supreme Court decision in *United States* v. *Halper,* suggest strongly that constitutional problems can result without such coordination.

3. Current trends in civil and criminal enforcement activities directed against organizations make the first two points particularly salient at this time.

4. The sentencing commission's development of organizational sentencing guidelines presents both an opportunity and a rationale for devoting greater attention to the relationship between these sanctions.

The report also looked at criminal and civil enforcement trends, noting that "the average collateral monetary sanction imposed on an organization far exceeds the average criminal fine."

Directors and Officers Liability

The basic legal duties of corporate directors are loyalty and care. That is, they are, first, to avoid conflicts of interest, and second, to be informed about company operations and not make poorly considered decisions or be negligent.

The corporate director and officer relationship to stockholders is similar to that of an agent to a principal. And, liability is similar in that it may be based on failure to perform a statutory or a common law duty. Failure to use ordinary care and prudence, when it results in loss, can generate liability.

Failure to use ordinary care and prudence, when it results in loss, can generate liability.

In 1985, a jury brought in murder verdicts against two former executives and a former plant foreman of Film Recovery Systems, Inc., of Elk Grove Village, Illinois. The three men were found responsible for the death of employees from arsenic poisoning; arsenic being used in the process of removing silver from film the plant processed. However, the appellate court overturned the murder and involuntary manslaughter convictions.

In another 1985 case, *Smith* v. *Van Gorkom,* the Delaware Supreme Court held that the 1980 sale of Trans Union Corporation to Pritzker's Marmon Group was "not the product of an informed business judgment" and that company directors were not completely candid with their stockholders. The court imposed individual liability on the directors of Trans Union and its chief executive, Jerome Van Gorkom, for failing to deliberate diligently enough about an offer to buy the company—they decided to sell the company after a two hour meeting and did not consult an investment banker. Two hours, the court said, was hardly enough time to make a reasoned business judgment. Trans Union's officers had done a great deal of research on the market and value of the company prior to the meeting and had concluded that the time was ripe for selling and that the offer from Pritzker was $18 above the then-market price of a Trans Union share—all of this was known to and conceded by the court. Still the Delaware Supreme Court delivered its negative opinion and tossed the case back to the lower courts to assess damages and awards, but Trans Union agreed to an out-of-court settlement of $23 million.

These two cases illustrate the extent to which improper conduct by executives and directors has been expanded far beyond previous legal definitions of wrongdoing and ordinary care.

The Rising Costs of Directors and Officers Suits

In 1995, the cost paid to the claimant for settlement averaged $4.53 million. The average legal defense for lawsuits against corporate directors and officers was $1.4 million. Shareholder suits continued to be the most common, representing 46 percent of the claims.

The study was done by the Chicago firm Wyatt Company, which has tracked directors and officers (D&O) claims and costs for 18 years. For the 1995 report, Wyatt surveyed 1,157 companies; 454 of the participant corporations had assets over $1 billion. The majority (65 percent) of survey participants were publicly traded corporations.

Inadequate or inaccurate disclosure or financial reporting was the most frequently cited general D&O claim issue. As stated, shareholders were the source of 46 percent of claims. Employment-related claims—job discrimination, wrongful termination, and harassment suits—are now 25 percent of all claims; the average cost of these claims is averaging more than $1 million.

Survey of Fiduciary Liability Claims

Another Wyatt Company survey, this one on fiduciary liability, gathered responses from 948 corporations during 1993 and 1994. More than 17 percent of the survey participants reported that their employee benefit plan assets exceeded $500 million, while the median plan asset size was about $50 million. The median number of employees covered by the plan was about 5,000.

The liabilities of benefit plan fiduciaries are spelled out in the Employment Retirement Income Security Act of 1974 (ERISA). Since most D&O policies exclude ERISA claims, insurance coverage is available, in a limited way, under fidelity bonds or employee benefit liability insurance.

The frequency of fiduciary liability claims has almost tripled since Wyatt's last survey, conducted in 1987. The costs of claims alleging violations of ERISA—the cost of a settlement or court award—is now $875,000. Additionally, closed claims with positive defense expenses had an average cost of just over $400,000, dramatically higher than a similar average defense cost of about $70,000 reported in Wyatt's 1987 study.

Benefits disputes accounted for about 44 percent of all fiduciary liability claims. Also reported were claims based on administrative error, which accounted for 10 percent. Plan participants brought 90.4 percent of the claims reported to the survey.

Conclusion

This discussion has touched on only the problem of expanding liabilities for organizations, officers and directors, and employees. It is important, however, to be aware of the constantly expanding legal threat. Remember, the landmark Van Gorkom decision, although called the worst decision in the history of corporate law, is now on the books as a legal precedent. And no congressional action has attempted to curb the power of federal prosecutors or modify any of the most expansive federal criminal statutes. Nor have the sanctions and fines on convicted organizations declined.

Checklists on Computer Protection, Foreseeability, and Negligence

Evidence of Foreseeability

	YES	NO
Serious crime in the area of the computer facility.	☐	☐
Knowledge of attacks on computer systems and how they occur.	☐	☐
Incidents of attempted and unauthorized access to your computer system.	☐	☐
Employees believe the e-mail system is similar to the telephone and use it for personal and business messages.	☐	☐
Your organization is highly dependent on computers.	☐	☐
Other computer facilities in your local area have had power or other environmental control problems.	☐	☐
Certain natural phenomena (floods, earthquakes, or tornadoes) are a threat in your geographic area.	☐	☐
You know that personnel are untrained or unaware of security precautions and procedures or disaster recovery duties.	☐	☐
The security staff has been given minimal training on possible computer security threats and appropriate countermeasures.	☐	☐

Notice

	YES	NO
Have you specific knowledge of a lack of or inadequate computer security in an area of your computing or communications system?	☐	☐
Did you receive this information via any of the following:		
1. written report, such as a loss report;	☐	☐
2. a security survey;	☐	☐
3. your insurer's survey and report;	☐	☐
4. a risk assessment;	☐	☐
5. a vulnerability analysis;	☐	☐

6. computer security penetration tests;   ☐ ☐
7. an internal audit;   ☐ ☐
8. a fraud hotline call;   ☐ ☐
9. a threat;   ☐ ☐
10. a warning.   ☐ ☐

Duty of Care

- Investigative notices.
- Take action; the level of protective action taken depends on risk foreseen.
- Cost-benefit analysis of protection systems and actions.

Questionnaire: Is Your Organization Negligence-Prone?

The following questionnaire highlights several areas that are basic elements of computer protection. A majority of "No" answers is an indication that your organization could face potential negligence and liability risks.

Management Attitudes and Actions

	YES	NO
Does management have a healthy skepticism that all controls are working properly?	☐	☐
Does management state that internal controls and monitoring are part of its responsibility in protecting the organization's assets?	☐	☐
Does your organization have an organization chart with management responsibilities clearly defined?	☐	☐
Do policies and procedures describe levels of responsibilities and approvals?	☐	☐
Is there an ongoing awareness of employee morale, employee attitudes, values, and job satisfaction level?	☐	☐
Is there a solid awareness that failure to protect the organization's assets could result in legal liability?	☐	☐
Are there adequate, legally maintainable standards of recruitment and selection?	☐	☐

Are measures taken to screen applicants for sensitive positions before appointment:

1. Employment verification?   ☐ ☐
2. Education verification?   ☐ ☐

3. Financial reliability?   ☐ ☐
4. Character?   ☐ ☐

Is there adequate orientation and training on computer protection and loss prevention matters and company policies with respect to sanctions for violations?   ☐ ☐

A disgruntled worker in the computer area has made repeated threats to harm or destroy the computer system—do you report this to management?   ☐ ☐

A staff member threatens to beat up or kill a coworker—again, do you know what to do?   ☐ ☐

Does the management information system provide for
  1. Monitoring operational performance levels for (a) variations from plans and standards; (b) deviations from accepted or mandated policies, procedures and practices; and (c) deviations from past quantitative relationships, that is, ratios, proportions, percentages, trends, past performance levels, indices, and so forth.   ☐ ☐
  2. Soliciting random feedback from or surveying customers, vendors, and suppliers for evidence of dissatisfaction, inefficiency, inconsistency with policies, corruption or dishonesty by employees.   ☐ ☐

Basic Financial Controls and Safeguards

Almost all fraud and embezzlements have occurred because specific controls were compromised, either intentionally or accidentally, or their warning was ignored by management.

                                                                 YES NO

Do your internal accounting controls include the following preventive measures:
  1. Separation of duties?   ☐ ☐
  2. Rotation of duties?   ☐ ☐
  3. Periodic internal audits and surprise inspections?   ☐ ☐
  4. Development and documentation of policies, procedures, systems, programs, and program modifications?   ☐ ☐
  5. Establishment of dual signature authorities, dollar authorization limits per signatory, expiration date, and check amount limits?   ☐ ☐
  6. Off-line entry controls and limits?   ☐ ☐
  7. Batch totals, hash totals?   ☐ ☐

Accounting and key personnel are secured with a fidelity bond. ☐ ☐

The purchasing department is separated from receiving responsibilities; the supervisor is not authorized to pay bills. ☐ ☐

The purchasing department uses prenumbered orders for all purchases; copies go to receiving and accounting. ☐ ☐

The purchasing department is audited periodically. ☐ ☐

Payroll checks are not signed by the person who prepared the payroll. ☐ ☐

The amounts paid are confirmed with the payroll records. ☐ ☐

Does a computer operator or bookkeeper handle cash, open incoming mail, mail statements, do follow-ups on delinquent receivables, approve write-offs of bad accounts, or approve refunds or credits? ☐ ☐

Is anyone responsible for authorizing modifications or "patches" to accounting or bookkeeping software programs? ☐ ☐

Are duties separated as between those with property-handling responsibilities and those with property-recording responsibilities? ☐ ☐

Are data processing validity checks made to determine that (1) a purchase has been approved by someone with authority to commit funds for such purposes, (2) from a vendor who is approved, (3) by a person who is authorized to buy, (4) that the specific goods ordered in fact were received, and (5) that the unit price charged and extensions are stated correctly on the vendor's invoice? ☐ ☐

Do controls and checks provide an oversight mechanism at each step in the processing of transactions that detects errors, omissions, and improprieties in the previous step (division of labor and dual responsibility for related transactions, such as a countersignature, segregation of functions, dollar authorization limits; thus forcing collusion by at least two parties to effect a fraudulent transaction)? ☐ ☐

Physical Security

	YES	NO
Is there an access control and physical security system?	☐	☐
Is the system totally functional?	☐	☐

	YES	NO
Are there security guards at the site?	☐	☐
Are all packages and briefcases searched going in and out?	☐	☐
Is positive ID required for all personnel entering the computer room or data center?	☐	☐
Are visitors controlled?	☐	☐
Is there a security log for visitors and employees?	☐	☐
Is access to the data center limited to working hours or office shift hours?	☐	☐
Is video used? If yes, in conjunction with tape?	☐	☐
Are internal fire prevention inspections conducted regularly?	☐	☐
Is the fire suppression system for the computer room tested regularly?	☐	☐
Does the computer center have fire retardant walls?	☐	☐
Is the computer room located under a large water system, company cafeteria, or bathrooms?	☐	☐
Do computers have an uninterruptible power supply and surge protection equipment?	☐	☐

Information Privacy and Confidentiality

	YES	NO
Does your organization have a written and distributed policy on the use of confidential third-party information or records?	☐	☐
Are employees required to read and sign a confidentiality agreement regarding information privacy?	☐	☐

Data Security

	YES	NO
Is sensitive and vital software and documentation secure (payroll records, personnel records, accounts receivable information, etc.)?	☐	☐
Is a backup file kept at a secondary site?	☐	☐
If yes, are any controls taken to protect it?	☐	☐
Do data backup and records retention and maintenance follow both laws and company policy?	☐	☐
Is there a thorough, appropriate, and tested disaster recovery plan?	☐	☐
Are restart procedures fully documented?	☐	☐

Are the resources shared with another company? ☐ ☐

Is there a log on modem access? ☐ ☐

Is any information encrypted? ☐ ☐

Computer Access Controls

YES NO

Are standard log-on procedures enforced through

1. Identification defenses, such as key or card systems; passwords or codes, alpha and numeric characters, with a minimum of seven characters; exclusion (repeated error lockout); time activator/deactivator; periodic code and password changes? ☐ ☐

2. Authentication defenses, such as random personal data, biometric identification, callbacks? ☐ ☐

Does someone have administrative responsibility for access authorization? ☐ ☐

Are system-stored passwords and codes encrypted? ☐ ☐

Is the sharing and disclosure of passwords allowed? ☐ ☐

Are all accesses logged? ☐ ☐

Are the date and time of access logged? ☐ ☐

Is establishment of access authorizations by levels of authority? ☐ ☐

Is establishment of authorizations by levels of security? ☐ ☐

Are the functions performed identified? ☐ ☐

Is the microcomputer or terminal identified? ☐ ☐

Are security violations logged? ☐ ☐

Are there exceptions logging systems that will detect

1. Out of sequence, out of priority, and aborted runs and entries? ☐ ☐

2. Out of pattern transactions: too high, too low, too many, too often, too few, unusual file access (odd times and odd places)? ☐ ☐

3. Attempted access beyond authorization level? ☐ ☐

4. Repeated attempts to gain access improperly—wrong password, entry code, or the like? ☐ ☐

5. Parity and redundancy checks? ☐ ☐

Are employees' passwords or access code privileges canceled immediately when employment terminates? ☐ ☐

Does the operator have restricted access to all programs and
data files in the mainframe?   ☐ ☐

Are critical or sensitive data files protected from unauthorized
access?   ☐ ☐

Are critical or sensitive data files protected from unauthorized
update?   ☐ ☐

Are employees allowed to load unauthorized software onto
the company's computers?   ☐ ☐

Are any operating or processing controls in place to detect
fraudulent manipulation of data?
1. Does software have change controllers that limit access
   to critical or specified files?   ☐ ☐
2. Do edit programs bar mainframe data that has been
   changed?   ☐ ☐
3. Is data tagged with the time and date of creation code? ☐ ☐
4. Is simultaneous access to a file or data field barred? ☐ ☐
5. Does the software provide an audit trail?   ☐ ☐

Does the operator maintain a record of what jobs are
processed?   ☐ ☐

Are daily transactions summarized?   ☐ ☐

Are master file balances summarized?   ☐ ☐

Are exception reports generated?   ☐ ☐

Is there a transaction log?   ☐ ☐

Is the system easily modified?   ☐ ☐

Are there restrictions on who has access to production copies
of live data or programs?   ☐ ☐

Is your computer network protected by firewalls or other
technology to control access?   ☐ ☐

Are dial-up lines monitored and recorded for repeated failed
access attempts?   ☐ ☐

Are standard mainframe access control measures employed
once dial-up connection has been made?   ☐ ☐

Does the network or data communications system use
encryption?
1. Does the key management program have adequate
   security?   ☐ ☐
2. Does the system have message and user identification
   and authentication?   ☐ ☐

Are there established and enforced secure procedures on how wire funds transfers are initiated and payment orders authorized?   □ □

Is a designated person allowed to send a funds transfer?   □ □

Are there identification and verification methods for the sender and the payment order?   □ □

Do wire transfer security measures include
1. Imposing limits on the amount of any transfer?   □ □
2. Transfers payable only from an authorized account?   □ □
3. Prohibiting any transfer that exceeds specific credit limits or account balances?   □ □
4. Limiting transfer to authorized beneficiaries?   □ □

2

Reducing Liability Risks
Compliance Programs

In a September 1996 decision, the Delaware Chancery Court said effective compliance programs could shield directors from liability for the wrongful acts of company managers and employees.

A basic legal duty of corporate directors and officers is to be informed about company operations and not make poorly considered decisions or be negligent.

The case involved the directors of Caremark International Inc., who were sued by the shareholders, on behalf of the company, for exposing Caremark to criminal liability. Caremark pled guilty in 1994 to making illegal payments to doctors to get them to prescribe Caremark's services. The suit charged the directors of Caremark with failure to supervise company personnel.

Caremark had an integrity-based compliance program that conformed to the Federal Sentencing Guidelines. The program had active participation by directors. The court ruling acknowledged this participation and the difficulty of establishing a completely effective system for reporting every illegal act or for ensuring the corporation would not violate any law or regulation. The court said report systems should be expected to work well enough to allow directors to make informed judgments about corporate compliance.

Because so many corporations are registered in Delaware, this decision should give another incentive to corporations to create effective compliance programs.

Organizational Due Diligence

The U.S. Sentencing Commission guidelines have given a broad legal definition and a blueprint for the establishment of "an effective program to prevent and detect violations of law." Having an effective compliance program means that the organization has exercised due diligence. The organization must, however, take a number of concrete and workable actions to demonstrate due diligence. The carrot of due diligence can mean a reduction in fines (the stick) for organizations convicted of a violation of law. This area is discussed in detail in the chapter on sentencing guidelines for individuals and organizations.

The following steps should be considered the minimum for an effective compliance program and used only as an outline guide.

Steps in Creating an Effective Compliance Program

1. Establish "compliance standards and procedures . . . reasonably capable of reducing the prospect of criminal conduct." This means written ethics policies and codes of conduct that discourage and deter unethical and illegal behavior. The codes should be distributed to management and employees and contain specific prohibitions.

2. Oversight of the compliance program must have "specific individual(s) within high-level personnel of the organization" in charge; that is, a senior manager, legal counsel, or a compliance officer. This person must be someone of high ethical stature. The organization must have "used due care not to delegate substantial discretionary authority" to persons the organization knew or should have known "had a propensity to engage in illegal activities."

3. There must be effective communication of organizational ethics policies and codes of conduct to all employees and agents. This may be done by "requiring participation in training programs or by disseminating publications that explain in a practical manner what is required."

4. The organization must take reasonable steps via monitoring and auditing systems that will detect criminal conduct by its employees and agents. The guidelines imply that an organization should have a full range of safeguards and information systems controls that would detect and deter waste, fraud, and abuse of assets as well as informal mechanisms related to organizational structure and management controls. In decentralized companies, branches and subsidiaries would need similar controls and audits.

5. Controls and safeguards must be monitored and compliance audits that will detect criminal conduct must be conducted.

6. Organizations must create and publicize a reporting system for reporting criminal conduct within the organization that would allow employees to do so without fear of retribution. One way to handle this requirement is with a policy directive from corporate management that clarifies when to report criminal conduct, under what circumstances, and to whom. Another way would be to establish a fraud hotline. The major elements that have been found necessary for a successful hotline are these:

A clear statement of the hotline's mission and objectives;

A staff with interview skills and compliance program knowledge;

Controls to protect the confidentiality of callers;

Internal guidelines to evaluate and classify allegations received through calls or letters;

A policy that inquiries into the allegations are performed by independent and qualified personnel;

Procedures to monitor cases to assure they are being handled and resolved properly.

7. Disciplinary mechanisms must be established for violations of law as well as for the failure to detect an offense. The "form of discipline that will be appropriate will be case specific." This step implies that everyone will be aware just what is a violation of law and that it always is possible to assess individual liability in a complex organizational structure. Adequate discipline, obviously, can conflict with union rules and employment laws. Personnel policy manuals may have to state that illegal or unethical conduct is grounds for termination. However, any manager should consult with legal counsel before terminating someone's employment.

8. Once an offense has been discovered, even though not fully verified, the organization must take "all reasonable steps to respond appropriately to the offense." This means the organization must start an internal investigation of the incident, and it will be "allowed a reasonable period of time" to conduct it. An incident need not be reported to the "appropriate governmental authorities" if the organization "reasonably concluded, based on the information then available, that no offense had been committed."

In summary, an effective compliance program must be driven from the top down. It must be motivated by top management's understanding of the dangers of prosecution and conviction for illegal acts and by an awareness that a compliance program can offer possible benefits including limits on corporate liability and awards for punitive damages, reduction of criminal penalties, and the positive public relations of being viewed as a good corporate citizen.

The sentencing guidelines offer organizations flexibility in setting up a compliance program. This allows an organization to identify the acts it must prevent and focus its educational activities in those areas.

Finally, the compliance program must be active and ongoing; it cannot be static. Policies, codes, controls, audits, investigations, enforcement, and responses must be monitored, reviewed, and updated in light of new legal developments and organizational experiences.

Ethics Policies and Codes of Conduct

An organization's basic legal defense strategy should include written ethics policies and codes of conduct that discourage and deter unethical and illegal behavior. The codes should be distributed to employees and contain specific prohibitions. Such policies and codes can reinforce legal norms within the organization and have a positive effect in deterring unlawful behavior. Unethical and illegal behavior must be clearly and unambiguously presented. Codes must be enforced, with enforcement procedures spelled out so that violations get reported, investigated, and disposed of.

Internal Controls and Safeguards

Organizations face new responsibilities for protecting assets and for reporting on the effectiveness of internal controls. In addition, it is critical that the relationship between internal controls and compliance programs be fully understood. As described already, internal controls are required for an effective compliance program, but no concrete definition was put forth by the sentencing commission guidelines. Some guidance, however, is found in the following laws and auditing literature.

The Foreign Corrupt Practices Act's Internal Accounting Control Standards

When the Foreign Corrupt Practices Act (FCPA) was enacted in 1977, the emphasis was on preventing corrupt payments to foreign officials. The law's antibribery section prohibits SEC-registered American businesses from certain corrupt practices in dealing with foreign officials. To determine if any bribery payments were made and to prevent companies from hiding them, the law imposes certain record keeping and internal control standards.

The FCPA has been used to prosecute domestic and foreign violations. The accounting standards section has become the most significant part of the FCPA because it makes it a criminal offense not to maintain accurate books and records and systems of internal controls.

The FCPA set no clear standard of wrongful intent for culpable conduct. The Omnibus Trade and Competitiveness Act of 1988 (PL 100-418) sets out standards for permissible conduct under the FCPA and contains increased statutory penalties for violations. It amends the FCPA by

1. Defining the level of culpability as "knowing"—That is, the person either knows, or has a firm belief, that someone is engaging in misconduct, that the circumstances exist, or that such result is substantially certain to occur. The prosecution must prove beyond a reasonable doubt that the defendant meets this requisite state of mind.

2. Adopting a standard of reasonableness—This means "a level of detail and degree of assurance as would satisfy prudent officials" that accounting controls and records are adequately maintained.

A criminal violation occurs when someone knowingly falsifies accounting records or circumvents or fails to implement an internal accounting control system. The FCPA amendments repeal the Eckhardt amendment, which provided a safe harbor from prosecution unless the organization also was found to have violated the act. Now individuals may face criminal liability regardless of whether the organization has been prosecuted.

Banking Law Requires Internal Controls

The Federal Deposit Insurance Corporation (FDIC) Improvement Act of 1991 requires issuance of an annual report by the largest depository institutions, signed off by management and an independent public accountant, on internal controls over financial reporting. The CEO, chief accountant, internal auditor, or financial officer must sign a report stating management's responsibilities for preparing financial statements and for establishing a yearly assessment of "the effectiveness of such internal control structures and procedures."

A Comprehensive Description of Internal Controls

A report entitled "Internal Control—Integrated Framework," issued in 1992 by the Committee of Sponsoring Organizations (COSO), developed integrated guidance on internal control. The guidance provided in this report is designed to give a common definition of internal control and establish a standard by which companies can assess their system of internal control.

Every business decision involves judgment and chance. And every system of internal control has its limitations, as described in the report. As such, fraudulent financial reporting and accounting surprises never can be eliminated totally. However, it is the COSO's view that, if management, boards of directors, external auditors, and others adopt the components, criteria, and guidelines set forth in the report, the incidence of such events will be reduced.

The report tries to settle on a definition of internal control in the hope that laws and regulations will adopt an agreed upon nomenclature. The study defines internal control as a process, "effected by an entity's board of directors, management and other personnel, designed to provide reasonable assurance regarding the achievement of objectives in the following categories:

- Effectiveness and efficiency of operations.
- Reliability of financial reporting.
- Compliance with applicable laws and regulations."

The study points out five interrelated components of internal control that must be present and functioning to have an effective control system:

- *Control environment*—This is the integrity, ethical values, and competence of the entity's people; management's philosophy and operating style; the way management assigns authority and responsibility and organizes and develops its people; and the attention and direction provided by the board of directors.

- *Risk assessment*—This is the identification and analysis of relevant risks to achievement of the objectives, forming a basis for determining how risks should be managed.
- *Control activities*—These are the policies and procedures that help ensure management directives are carried out. They include a range of activities as diverse as approvals, authorizations, verifications, reconciliations, reviews of operating performance, security of assets, and segregation of duties.
- *Information and communication*—These include generated data as well as information about external events plus clear messages from top management that control responsibilities must be taken seriously.
- *Monitoring*—This process assesses the quality of the system's performance over time.

The Auditor as Fraud Detective

On December 22, 1995, the Private Securities Litigation Reform Act (PL 104-67) became law. The act amends the Securities and Exchange Act of 1934 (15 U.S.C.S. 78a et seq.) and requires auditors to design audit procedures that will likely detect illegal acts that would have "a direct and material effect" on a publicly held company's financial statement.

With computers often the main tool used to create results used in financial statements, auditors must look at computer safeguards and controls as well as the overall internal controls of a company. Auditors also must assess the risks and vulnerabily of a company to fraud. If illegal acts are thought to be possible or are discovered, these must be reported to top management or the audit committee. If the auditor's report does not receive timely action by management, the auditors must inform the Securities and Exchange Commission.

This new law may lead companies to take a more proactive stance toward strengthening internal controls, security, and compliance programs.

Policy Guide: The Example of Internal Theft

The U.S. Sentencing Commission guidelines offer a blueprint for an effective and comprehensive program to prevent and detect violations of law. Although the guidelines are meant for a host of federal crimes, theft, usually a state offense, easily can become federal; for example, if the mail or telephone are used or if a state line is crossed.

Developing an antitheft policy often is difficult because employee theft is complex. One reason for the difficulty is that it is common for theft and pilferage to be widespread in an organization, involving employees in any department from maintenance, food service, offices, and data processing. And the type of items stolen have a wide range of value. These are merely a couple of problems encountered when trying to come up with a companywide antitheft policy.

In attempting to reduce theft, an important task of management is to express clearly to employees that theft is considered to be and will be treated as a serious problem. The major consequence of a theft policy is that it conveys to employees the organization's concern for theft. Corporate officials can unambiguously demonstrate that theft of company property will invoke sanctions on the employee who steals.

At the same time, management must be aware of workplace privacy issues. Policies must cover inspections and searches of company equipment and facilities used by employees, employee surveillance, investigative interviews and reports, and disciplinary measures. Each of these areas is a legal trap that could lead to charges of malicious prosecution, invasion of privacy, defamation, false imprisonment, or infliction of emotional distress.

Antitheft Policy Considerations

We did an informal survey of organizations that had experience in court or arbitration with employees charged with theft.

The most consistent finding was that judges or arbitrators first asked if the organization had a policy covering employee theft and was such a policy organizationwide, did it cover all employees, and was it issued by top management.

Judges and arbitrators also looked critically at the following:

1. Was the policy developed because the organization had past experience of serious losses due to employee theft and pilferage? If the organization stated that it had had serious losses, judges often asked if this could be substantiated with internal audit or security reports or insurance claims.

2. Did the policy or company rules state a dollar value threshold for items stolen? Judges seldom asked this question, although it was often implied. That is, some judges tended to get angry with an organization that charged or dismissed employees for stealing low value items.

3. Did the policy spell out precisely what the company would do to an employee stealing company property? This question was almost always asked. If the company antitheft policy had an immediate termination or no exceptions clause, the company had to be prepared to defend it. Even though such a clause may be consistent, most companies prefer a two- or three-stage procedure, moving into a disciplinary policy of warning, counseling and reminder of the policy, and then termination or prosecution. The company policy concerning prosecution always should be clearly stated and be in the name of the chief executive officer. The prosecution policy should include a requirement that anticipated prosecutive actions be reviewed with legal counsel prior to their initiation.

4. Did the employee know the antitheft policy rules and did the employee acknowledge this? This is another question always asked when the organization has a policy. Many companies have antitheft policies inserted in the Employee Handbook and employees do a one-time sign-off on the entire handbook. However, the courts are moving in the direction of having company rules and policies clearly and repeatedly communicated to employees and consistently applying those rules.

In the following case, a firing was upheld partly because company policy was communicated to the employee. An employee of an auto parts store took home a software package and a cash register procedure manual without getting the company's permission. When the loss was discovered and traced to the employee, he was first suspended, then after a management review, he was fired.

The employee and his union filed a grievance and the case went to arbitration.

The auto parts company said it was justified in firing the employee because company losses to internal theft were running at $2 million annually, it had a strict security program, it had a policy of consistently terminating employment of any employee who took company property without permission, and all employees knew this policy and the termination procedures.

Further, this employee, according to the company, once had held a management position where he was responsible for both communication and enforcement of antitheft rules. Also, the employee recently had been reprimanded for violating the rules and reminded of the rules and his duty to follow them.

The employee admitted he took the materials without company permission, but he argued that he had removed the materials so he could learn more about the computer system and that they were related to his job. Also, these particular materials were not part of the inventory shrinkage problems affecting the company.

The arbitrator's opinion stressed that the employee knew he needed permission to remove the software package and the consequences of doing so. The arbitrator said the company had applied its policy and procedures stringently and consistently, with dismissal or prosecution as the ultimate penalties. The arbitrator upheld the firing (in the Matter of East Bay Automotive Machinists Lodge No. 1546 and Grand Auto, Inc., California Arbitration Award, Case No. 04-30-93).

3

Protecting Intellectual Property Rights

*I*ntellectual property generally refers to ideas and information residing in various formats that is given the legal status and protection of property as an asset of an individual or enterprise.

Article I, Section 8, clause 8 of the Constitution authorizes the Congress to establish intellectual property law: "Congress shall have Power . . . To promote the Progress of Science and useful Arts, by securing for limited Times to Authors and Inventors the exclusive Right to their respective Writings and Discoveries."

Patents, trademarks, and copyrights are the three primary forms of intellectual property rights in worldwide use. They encourage the introduction of innovative products and creative works to the public by guaranteeing their originators a limited exclusive right, usually for a specified period of time, to whatever economic reward the market may provide for their creations. Other types of intellectual property rights include trade secrets, "mask works" (i.e., the pattern on the surface of a semiconductor chip), and industrial designs.

The chapters that follow detail U.S. intellectual property law as well as reference international agreements. First, though, we will look at several recent estimates of the extent of theft and loss of intellectual property.

Intellectual Property Theft

The American Society for Industrial Security (ASIS), in its "1995/96 Trends in Intellectual Property Loss Special Report," found

- An increase of 323 percent in the number of intellectual property loss incidents reported per month, from an average of 9.9 incidents in 1992 to an average of 32 incidents in 1995;
- Potential losses from intellectual property theft for U.S.-based companies are estimated to be $24 billion annually;
- Nearly 47 percent of all foreign incidents occurred in England, Canada, and Germany;
- Top five nationalities involved in incidents were Chinese, Canadian, French, Indian, and Japanese;
- Insiders with a trusted relationship were involved in nearly 75 percent of all incidents;
- Of the reporting companies, 76 percent have formal safeguarding proprietary information programs.

Loss of strategic plans, R&D, and manufacturing process information accounted for over 60 percent of the financial losses.

William J. Cook, an intellectual property lawyer, in testimony in February 1966 before the House Judiciary Committee's Courts and Intellectual Property Subcommittee, said, "today, the Internet connects 49 million people in 96 countries. This means that the value of intellectual property developed over months or years can be destroyed by worldwide dissemination on the Internet in a matter of days or hours . . . A million dollar computer program released on Monday, improperly obtained on Tuesday, and uploaded to the Internet on Wednesday, becomes shareware by Thursday and worthless freeware by Friday."

Cook proposed the enactment of legislation that would require Internet service providers to remove copyrighted material from their networks as soon as they learn it has been posted.

Cook said, "Service providers are uniquely well-positioned to perform this function and limit online infringement. They can remove infringing material with much greater speed than any court procedure. And a speedy removal is essential to protecting the value of copyrighted material posted online."

Urging Congress to take the initiative, Cook said the rapid expansion of on-line copyright infringement on the Internet seriously threatens the value and future of copyrights.

Copyright Piracy Costs Grows

In 1982, the cost of piracy to U.S. companies was approximately $5.5 billion, according to the International Intellectual Property Alliance. By 1995, it was estimated to be $200 billion. For the computer software industry, estimates are that two out of five software programs sold are counterfeit; this would mean that sales of counterfeit software almost exceed 40 percent of total industry revenues.

In May 1996, the Clinton administration said that China was not satisfactorily implementing the Agreement on Enforcement of Intellectual Property Rights and Market Access, signed in March 1995. The administration said that "critical deficiencies were present in China's implementation of measures to address piracy at the production and wholesale distribution level . . . Due to lax enforcement at the point of production and at the border, exports of pirated computer software . . . and other products have grown substantially over the past year."

The International Intellectual Property Alliance claims Chinese piracy of entertainment materials and computer software totaled $1.2 billion in 1995, up from an estimated $866 million in 1994.

Threatened with trade sanctions from the United States, China and the United States reached an agreement under which China would close certain factories known to be producing pirated materials. Those violating intellectual property rights now will be subject to criminal prosecution. Any Chinese factory producing compact disks must first get approval from its National Copyright Administration, which can check with the copyright holder for approval.

The final agreement did not contain some gain of market access to China by U.S. firms. And China will not allow any thorough on-site inspection of factories by U.S. officials to see if pirating-related equipment and products have been destroyed. Instead, local Chinese officials will keep records of audits showing enforcement of the agreement.

China certainly is not the only violator of intellectual property. In Europe, it is estimated that half the software in use is pirated. Yet, the piracy rate is even higher elsewhere, according to the Business Software Alliance. A 1994 survey found piracy in Japan was 67 percent; in Taiwan, 72 percent; and in South Korea, 78 percent. The software piracy rate in the United States was estimated to be 35 percent.

International Agreements to Help Curb Intellectual Property Theft

While trade groups use education as a primary tool to help lower piracy rates and governments threaten trade sanctions, there are important international agreements on intellectual property rights.

The Agreement on Trade-Related Aspects of Intellectual Property Rights (TRIPS Agreement) of the Uruguay Round of Multilateral

Trade Negotiations under the General Agreement on Tariffs and Trade (GATT) took effect for the United States on January 1, 1995.

The legislation strengthens the GATT by replacing the current system of settling trade disputes between member countries with a World Trade Organization (WTO). The WTO would have the power to monitor member countries' implementation of trade commitments and settle trade disputes by consensus. Under the WTO, consensus means that a ruling would go into effect unless all member countries opposed a particular ruling.

Each GATT member must give nationals of other member countries treatment no less favorable than it gives to its own nationals regarding the protection, maintenance, and enforcement of intellectual property rights. Also, any advantage, favor, privilege, or immunity granted by a GATT member to the nationals of any other country shall be accorded immediately and unconditionally to the nationals of all other GATT member countries.

For intellectual property, the GATT agreement also provides

- Improved standards for protecting intellectual property;
- Better enforcement of rights, both internally and at borders;
- Dispute settlement options including trade sanctions and increased tariffs for noncompliance.

GATT is seen by many critics and supporters as offering a baseline of protection and as but one element in the ongoing U.S. effort to improve worldwide protection of intellectual property rights.

Throughout this discussion, specific provisions in the GATT/TRIPS agreement often will be compared with the North American Free Trade Agreement (NAFTA), which contains provisions that offer stronger protection for U.S. intellectual property.

The agreement says that computer programs and databases are protected subject matter under copyright and that computer programs are literary works as stated in the Berne Convention. The Berne Convention for the Protection of Literary and Artistic Works (1971), Article 2, covers "new" expressive works or forms of "writing" that include computer programs.

In GATT, "computer programs shall be protected as literary works (Sect. 1 Art. 10.1), and compilations of data [by virtue of the selection or arrangement of their content] shall be recognized as intellectual creations and shall thereby receive copyright protection" (Sect. 1 Art. 10.2). Article 9(2) protects only the "expression," rather than ideas, procedures, methods of operation, or mathematical concepts; this also is found in Section 102(b) of the U.S. Copyright Act (17 USC).

The Berne provisions on "moral rights" were not incorporated into the agreement.

Copyright holders in computer programs are given the right to authorize or prohibit rental of their products.

A minimum term of protection of life of the author plus 50 years or 50 years from first authorized publication is set for copyright protection (Sect. 1 Arts. 12 and 14).

Not covered or defined is a right of "public" communication; a broad and commercially meaningful definition is critically important given the increase in distribution of copyrighted materials over electronic networks.

NAFTA covers "the right to authorize or prohibit the communication of a work to the public" (Art. 1705, 2(c)).

Enforcement of Intellectual Property Rights

Members of GATT must ensure that "enforcement procedures . . . are available under their national laws so as to permit effective action against any act of infringement of intellectual property rights covered by this Agreement, including expeditious remedies to prevent infringements and remedies which constitute a deterrent to further infringements" (Sect. 1 Art. 41.1). Injunctive relief and damages provisions are in Section 1, Articles 44 and 45.

Enforcement provisions mandate the imposition of deterrent criminal penalties against willful trademark counterfeiting or copyright piracy. The agreement calls for "imprisonment and/or monetary fines sufficient to provide a deterrent."

Nullification and impairment provisions provide a multilateral procedure to redress violations of the spirit of an intellectual property agreement. This means, for example, a WTO member could be taken to a WTO dispute panel for failing to permit the transfer of royalties from the commercial exploitation of an intellectual property right.

Legal protections are only as adequate as the willingness of the GATT member countries to enforce that protection. However, enforcement measures and procedures should be monitored so that they do not become barriers to legitimate trade.

Compliance Schedules

GATT countries have varying time periods in which to comply with the obligations of the agreement:

- Developed country implementation was July 1, 1996.
- Developing countries and countries that are in the process of evolving from a centrally planned to a free enterprise economy have up to five years; an end of the "nullification and impairment" moratorium.
- Least developed countries are given 11 years.

Japan, Taiwan, and Korea are fully developed countries and subject to the 1996 deadline. It is felt, however, that these transition periods are too long for developing and least-developed countries. U.S. industries will be delayed in reaping full commercial benefits in the developing countries, as they are not obligated to provide protection.

Improvements Within and Outside of GATT

Specific suggestions to strengthen the GATT and U.S. intellectual property rights worldwide include the following:

- Active and strong leadership of the United States in the World Trade Organization and the intellectual property council.
- Bilateral negotiations to gain improved intellectual property protection in individual countries, with the provisions of the intellectual property rights in the NAFTA as the starting point.
- Protection against parallel imports (as in the U.S. Copyright Act of 1976, 17 U.S.C. 602).
- Full national treatment to all intellectual property rights holders, which is to thwart the oft-used ruse of countries creating "new" subject matter, or rights, or "new" beneficiaries of new or old rights, and then arguing that these are outside the realm of existing copyright and "neighboring rights" conventions. This, then, frees these countries from any obligation to respect and provide national treatment to foreign intellectual property owners.
- Use of "Special 301" sanctions (which include tariffs and fines) to remedy trade discrimination and protect U.S. intellectual property rights in developing countries during the transition periods set out in the agreement.
- Adoption of a "negligence" standard of "fair and reasonable degree of care" for third-party acquisition of trade secrets.

4

Trade Secrets

Information permeates an enterprise; it is often elusive, defying a flowchart or attempts at "mapping" its locations. Proprietary information is less slippery, having an importance attached to it by management. Intellectual property has legal status and therefore a higher recognition. But not all important information, having value, is necessarily given this recognition at its birth. Much of it is in a state of becoming, not yet given the status of a proprietary or an intellectual property niche. When information reaches a stage where it is seen as having value or potential value, as an intangible asset that can contribute to the enterprise's earning power, it deserves protection.

Trade secrets are information that becomes wrapped in concepts of "value" and "protection" as well as "secrecy."

The essence of a trade secret is that it has value; its ultimate value occurs when it finds the highest and best use, yielding the greatest return through exploitation, either within the enterprise or by transferring rights to others via licensing and royalty arrangements. Assigning a final value to trade secret information may be impossible given the many, often subtle, permutations of valuation. One thing is sure: an unused trade secret could be called worthless.

Before it can be exploitable, trade secret information is in a stage of becoming, the potential is visible, but the "package" is not ready. At this stage, however, the element of protection must enter, along with some record keeping and bookkeeping and legal matters. Record keeping must start with identifying the information to be protected, retaining all the related documentation during development, and making

39

sure all documents are signed and dated. Bookkeeping should be not only a record of costs (time, effort, and expense) of information development but the costs of protection as well. And all records should be created and maintained to meet the business records standard under the rules of evidence.

The protection and value of trade secret information are joined in a series of cost-benefit analyses. This implies the use of periodic, focused audits of trade secret programs to monitor development, value, and protection compliance.

To understand trade secrets we must first look at the relevant statutes and case law.

Trade Secrets Law

Technological and business information that is used secretly within an enterprise, that lends a competitive advantage, and that is not known generally by competitors is legally protectable as a trade secret. Matter protectable as a trade secret is broad and includes varieties of information for which patent protection is never available. For many types of technological information, such as complex industrial processes and formulas, trade secrets and patents are alternative forms of protection. Other innovative matter, such as computer software and related developments, which may be marginally protectable by patent or copyright, are better protected as trade secrets.

Trade secret law has been ruled by the Supreme Court to be independent of and complimentary with the patent system. This allows, for example, the choice of either seeking patent protection for computer software or retaining the matter as a trade secret (see *Gottschalk* v. *Benson*, 409 U.S. 63 [1972]).

Trade secret law generally is based on common law and contractual provisions. Whereas patent and copyright are under federal law, the law of each state defines what constitutes a trade secret, the rights of the holder, and the enforcement of all trade secret claims. State trade secret law is not preempted by federal law. And, unlike patent or copyright, there is no limit on the duration of a trade secret.

A federal statute, the Trade Secrets Act (18 U.S.C. 1905), prohibits unauthorized release of any information relating to trade secrets or confidential business information by a federal officer or employee.

Elements of Trade Secrets Law

For confidential information to be given the status of a trade secret it must be commercial information. That is, it cannot be just any information that a firm does not want known, such as an internal report that discloses poor management practices. Also, the commercial information must have a value that lies in the competitive advantage it gives over business rivals. Another essential element of a trade secret is its confidential nature, which must be maintained.

In 1979 the Uniform Trade Secrets Act (UTSA) was approved by the National Conference of Commissioners on Uniform State Laws and amended in 1985. The UTSA, which has been adopted, with variations, into the civil codes of over 30 states, gives the following definition of a trade secret:

> "Trade secret" means information, including a formula, pattern, compilation, program, device, method, technique, or process that:
>
> (i) derives independent economic value, actual or potential, from not being generally known to, and not being readily ascertainable by proper means, by other persons who can obtain economic value from its disclosure or use, and
>
> (ii) is the subject of efforts that are reasonable under the circumstances to maintain its secrecy.

A trade secret takes on the attribute of property to be protected, as another asset of the firm. In the *Restatement Second of Torts* (1965), Section 757, we find further clarification of factors to be considered in determining whether information can be a secret:

1. The extent to which the information is known outside of the business;
2. The extent to which it is known by employees and others involved in the business;
3. The extent of measures taken by the business to guard the secrecy of the information;
4. The value of the information to the business and its competitors;
5. The amount of effort or money expended by the business in developing the information; and,
6. The ease or difficulty with which the information could be properly acquired or duplicated by others.

To establish a claim of misappropriation of a trade secret, the restatement says:

1. There must be a protectable interest, i.e., a trade secret;
2. The plaintiff must have a proprietary interest in the trade secret;

3. The trade secret must be disclosed to the defendant in confidence or it must be wrongfully acquired by the defendant through improper means;
4. There must be a duty not to use or disclose the information; and,
5. There must be a likely or past disclosure or use of the information, if in a different form, which is unfair or inequitable to the plaintiff.

The Concept of Novelty

To be protected as a trade secret, the information must be novel; that is, it cannot exist in the public domain. However, the information could be combined in such a way to make it new and uniquely different and possibly qualify for trade secret protection.

Information that could be a candidate for trade secret status often is created and stored on a computer. Take, for example, market research data. As it is collected in raw data, such as survey questionnaire tabulations, it may not meet the requirements of a trade secret but it would certainly justify being protected as proprietary information. However, the methodology used, say, a proprietary survey sampling technique or the mailing list, if it was compiled in-house using customer names, should qualify as a trade secret.

Then again, all of the elements used in a marketing research survey—the research methodology, the survey select sample, the statistical analysis—may be commonly known. Yet, the combination of these elements can result in marketing information or a plan that qualifies as a trade secret.

Two elements are now present: unique, novel information not in the public domain and information that confers a commercially competitive advantage or potential advantage to its owner. The novelty of the plan or one of its elements is not necessarily critical to trade secret status; it is the totality of the plan that is unique in that everything about it is not common industry knowledge and that the marketing plan confers a competitive advantage.

How Trade Secrets Can Lose Their Confidentiality

Proper means to discover a trade secret include independent invention, reverse engineering, observation of the item in public use or on public display, and obtaining the information from public literature.

The clearest way to abandon secrecy is to make public disclosure of the information. This does not necessarily mean broad public dissemination; a single third party will suffice as long as it is made in the absence of confidential circumstances. A patent issuance will end trade secret status. The age or relevance of the information can also affect its trade secret status.

The body of trade secret laws relevant to information systems has grown steadily. In one of the most prominent cases (*Telex Corp.* v. *IBM*), IBM recovered $23 million from Telex for theft of IBM product development material. There have been numerous prosecutions for theft of computer programs and software.

Secrecy and Protection

With trade secrets we find the critical interdependence of law and protection. In asserting trade secret ownership, one "has the burden of establishing that he took adequate safeguards to protect secrecy." Measures that have met previous legal tests of safeguards include employment agreements and practices of nondisclosure, vendor contracts, and the range of stringent security practices common in protecting classified information.

Not only must sufficient security measures be in place to ensure secrecy, the information owner should be prepared to provide detailed evidence that the security was "reasonably sufficient under the circumstances" should a trade secret issue land in court.

Security measures normally could include

1. Hiring and discharge policies and practices, preemployment clearance;

2. Secrecy agreements;

3. Need-to-know rules, compartmented use or documents legended with clearly marked security classifications covering hierarchies of secrecy (confidential, proprietary, trade secret, etc.);

4. Document-handling procedures, logging, storage, declassification, archiving, and destruction;

5. Risk and vulnerability analyses, physical security policies, procedures, measures, equipment and systems, and administration; access control covering visitors, vendors, and employees;

6. Trade secret or security awareness programs;

7. Audits of all legal, administrative, and security training and awareness programs related to trade secret protection; also regular limited-scope compliance audits;

8. Prepublication clearance for articles and information dissemination;

9. Enforcement and sanctions for violations of any security measures or disclosure of trade secrets.

The legal interpretation of the phrase *reasonable under the circumstances*, with regard to safeguarding trade secrets, has varied in the courts. In several cases, the courts have held that security need be only reasonable efforts. In a 1970 case (*E.I. duPont de Nemours & Co. v Christopher*), the court ruled that: "We should not require a person or corporation to take unreasonable precautions to prevent another from doing that which he ought not to do in the first place. Reasonable precautions against predatory eyes we may require, but an impenetrable fortress is an unreasonable requirement."

In a 1991 case (*Rockwell Graphics Systems* v. *DEV Industries, Inc.*), Rockwell Graphics accused DEV of misappropriating valuable design drawings. DEV countered by arguing that the drawings had been obtained lawfully and that Rockwell had given up its right to trade secret protection because the drawings were circulated widely to independent machine shops and because the company had not taken reasonable security measures to guard their secrecy. Rockwell said it had confidentiality agreements with the independent machine shops and that the cost of maintaining secrecy should not be unreasonable.

Rockwell also filed a federal RICO (Racketeer Influenced and Corrupt Organizations) lawsuit against DEV.

The case went to the U.S. Court of Appeals in Chicago after a federal district court agreed with DEV. Judge Richard Posner, writing for the appeals court that reversed the district court's verdict, said:

This is an important case because trade secret protection is an important part of intellectual property, a form of property that is of growing importance to the competitiveness of American industry . . .

If trade secrets are protected only if their owners take extravagant, productivity-impairing measures to maintain their secrecy, the incentive to invest resources in discovering more efficient methods of production will be reduced, and with it the amount of invention.

Posner said the courts should balance the costs of maintaining secrecy against its benefits in determining whether a company's security practices were reasonable enough to justify trade secret protections.

On the other hand, more than just good intentions and a show of security are necessary to claim adequate protection was afforded confidential information. The courts will look at each case involving misappropriation of alleged trade secrets and make a determination as to the effectiveness of security.

The courts have given guidance on misappropriating a trade secret. The statutory definition is that no person, including the state, may misappropriate or threaten to misappropriate a trade secret by the following:

a. acquiring the trade secret of another by means which the person knows or has reason to know constitute improper means;

b. disclosing or using without express or implied consent a trade secret of another if the person did any of the following:
   (i) used improper means to acquire knowledge of the trade secret;
   (ii) at the time of disclosure or use, knew or had reason to know that he or she obtained knowledge of the trade secret through any of the following means:
   a. deriving it from or through a person who utilized improper means to acquire it;
   b. acquiring it under circumstances giving rise to a duty to maintain its secrecy or limit its use;
   c. deriving it from or through a person who owed a duty to the person seeking relief to maintain its secrecy or limit its use;
   d. acquiring it by accident or mistake.

The phrase *improper means* has been defined to include espionage, theft, bribery, misrepresentation, and breach or inducement of a breach of duty to maintain secrecy.

Secrecy Agreements

A secrecy agreement is a way for a company to keep its technical or business information confidential. A secrecy agreement is an agreement that creates a confidential relationship with either an employee or someone outside the company, such as a supplier or subcontractor. It is used for preserving secrecy with respect to inventions, technical information, know-how, and other information that may qualify as trade secrets.

A trade secret or confidentiality clause should be part of agreements with employees, independent contractors or consultants, visitors, vendors, distributors, lenders, partners, and shareholders.

A signed secrecy agreement usually is necessary to create or preserve intellectual property protection rights. Secrecy agreements should not be entered into casually and without legal advice or assistance.

Even though there is a common law duty on employees not to disclose or use their current or former employer's trade secrets and a fiduciary duty for officers, a signed agreement is evidence of an employer's action to protect trade secrets.

Elements of a Secrecy Agreement

Secrecy agreements should cover the following:

1. The employee you want to enter into an agreement with;

2. The departments or subunits of the company involved;

3. Where the agreement is likely to be performed (that is, in which state of the United States?);

4. A definition of trade secrets and the trade secret information to which the employee has access (indicate the particular product(s) or manufacturing process to which the information relates as well as the kind of information, such as software, designs, complete sets of manufacturing drawings, quality assurance reports, marketing studies or data, test specifications, or process details);

5. When the information is to be available to the employee (if over a period of time, say, the length of a contract or project, give the period of time);

6. The fact that the company is protecting its trade secret information;

7. That the employee will not disclose or misappropriate trade secret information;

8. That the employee will report any unauthorized disclosure or use of trade secret information;

9. The period of time the employee is to keep the information in confidence (again, be specific; there should be notice of postemployment nondisclosure; this may be a separate agreement in the form of a restrictive covenant, which is a provision reasonably restricting—not too broad as to time, territory, or activity—competition by the employee after the employment is finished; as the enforcement of restrictive cove-

nants vary with states that permit them, they must be carefully drawn to conform with a state's law);

10. If action is taken against misappropriation by an employee and injunctive relief is sought, set an agreed amount of bond or security to protect the employee against whom the injunction may be issued.

Points to Remember About Secrecy Agreements

- Secrecy agreements must be reasonable in scope and not contrary to public policy.
- Only top management should be authorized to enter into and sign secrecy agreements made with others.
- Secrecy agreements are enforceable, and injunctions may be granted and large damages awarded for not adhering to them.
- Avoid disclosing trade secret information to any outsider except under an enforceable secrecy agreement.
- Consult with legal counsel before entering into any secrecy agreements.

Remedies for Misappropriation of Trade Secrets

Criminal Prosecution

State statutes covering trade secret theft usually contain criminal sanctions.

Injunctive Remedies

As soon as a trade secret misappropriation is discovered, the quickest legal tactic is to ask the court for a temporary restraining order or preliminary injunction. This forces the offender to immediately cease violating a confidentiality agreement.

You must be ready to show the court the value of the information and the damage the enterprise will suffer if the trade secret information is disseminated or misused.

Recovery for Damages

For trade secret misappropriation by employees compensatory damages can be recovered. In determining damages, the court may examine the trade secret owner's loss or the defendant's gain or both. Here again, knowing the fact of damage caused by the loss of a trade secret is the critical first step; the next step is to determine and show the amount of damages.

Punitive damages may also be awarded if the defendant's conduct was "willful and malicious misappropriation" of the owner's trade secret.

Federal Statutes to Counter Trade Secret Theft

Economic Espionage Act

The Economic Espionage Act of 1996 (PL 104-294) is designed to protect proprietary economic information and creates criminal penalties for the wrongful copying or control of trade secrets or the wrongful diversion of a trade secret to the economic benefit of someone other than its owners.

The act amends Title 18 of the U.S. Code by inserting a new chapter, 90. Section 1831 of the chapter prohibits anyone from obtaining trade secrets by fraud, theft, or deception for the benefit of a foreign agent or government. Individuals can be fined up to $500,000 and imprisoned for up to 15 years. Organizations can be fined up to $10 million.

Receiving, buying, or possessing a trade secret known to be stolen is an offense. Attempts to steal and conspiracies to steal trade secrets also are prohibited. A trade secret is defined as "all forms and types of financial, business, scientific, technical, economic, or engineering information" having independent economic value and for which the owner has "taken reasonable measures to keep such information secret."

Congress intended this act to help federal law enforcement groups combat trade secret thefts by foreign companies, often with the cooperation of foreign governments, and thefts by U.S. employees. Also, no federal criminal statute dealt directly with economic espionage or the protection of proprietary economic information. And, under existing law, no statutory procedure protected the victim's stolen information during criminal proceedings. In any prosecution or proceeding, the court may issue orders to preserve the confidentiality of trade secrets. In a civil action, the attorney general may issue injunctive relief against any violation under this act.

Under Section 1832, it is an offense to convert a trade secret that is related to or a part of a product in interstate or foreign commerce, if the trade secret was obtained without authorization or through theft. It also is unlawful for anyone, without authorization from the owner, to make copies of trade secret information, communicate such information, receive or buy, or to attempt or conspire to do any of these actions that would cause injury to the trade secret owner. Conviction carries a fine or imprisonment for not more than 10 years or both. Organizations can be fined up to $5 million.

The statute includes a criminal forfeiture provision in addition to any other sentence. Any property or proceeds derived from the crime or property used or intended to be used in the crime may be forfeited.

National Stolen Property Act

The National Stolen Property Act (18 U.S.C., Sect. 2314) calls for criminal sanctions against any person who "transports, transmits, or transfers in interstate or foreign commerce any goods, wares, merchandise, securities or money, of the value of $5,000 or more, knowing the same to have been stolen, converted or taken by fraud . . . " Penalties can be a fine of up to $10,000 or a prison sentence of 10 years or both.

Federal courts have ruled that confidential information stored on a computer was valuable property under the definition of "goods, wares, or merchandise"; and a person who "transmitted" stolen proprietary business information from one computer to another across state lines could be prosecuted under the statute.

The key clauses of the statute that must be satisfied for a conviction are these:

1. The items must be transported or transmitted in interstate or foreign commerce.
2. The items must meet the definition of goods, wares, merchandise, securities, or money.
3. The items—property or money—must have a value of $5,000 or more.
4. The defendant must have knowledge that the items were stolen or falsely made.
5. The defendant must know that the items were stolen, converted, or taken by fraudulent means.

The aims of the statute are to prohibit the use of interstate transportation facilities to move stolen goods and to punish theft of property that was beyond the capability of an individual state. Therefore, the "movement" of a trade secret across state lines must be established. Also, it must be established that the defendant had knowledge that the information was property and that it was stolen.

Obviously, other federal laws come into play when stolen information is transferred or transmitted. First among the laws are the wire or mail fraud statutes; these statutes often are merged with or underlay a charge of conspiracy.

International Protections for Trade Secrets

Protection of trade secrets (called *undisclosed information*) in all GATT countries is covered in Article 39 of that agreement. Natural as well as legal persons may prevent information being disclosed to, acquired by, or used by others without their consent so long as the information is relatively secret, has commercial value because it is secret, and has been the subject of reasonable efforts to keep the information secret.

NAFTA uses the broader phrase *"actual or potential" commercial value* (17, Article 1711(1)(b)).

A gross negligence standard is established for disclosure of trade secrets to third parties.

Conclusions

- Keep records of the time and investment expended to create trade secret information.
- Identify by name and position all persons to whom trade secret information was disclosed.
- Use risk and vulnerability analyses to determine the threat of loss of trade secret information.
- Implement physical security measures that are obvious, reasonable, and adequate.
- Keep records of the costs of protection.
- Document and periodically audit all protective measures.
- Remember that trade secret protection is an ongoing program of cost-benefit analysis; protection should be adequate to the threat and value.
- Confidentiality and secrecy agreements should be specific in terms of individual, information, location, and time or duration; a companywide trade secret statement likely will be inadequate.
- Proprietary confidential information is seldom static; technology and service product life cycles have shrunk, often to less than two years. Protective measures must be as dynamic as the confidential information.

5

Trademarks
and Patents

Trademarks are words, names, symbols, devices, or a combination of these used by manufacturers or merchants to identify their goods and distinguish them from others. Service marks perform the same function for services. *Trademark counterfeiting* generally refers to the deliberate, unauthorized duplication of another's trademark; *trademark infringement* refers to the unauthorized use of a trademark that is so similar to an existing trademark that, considering the products involved, consumers are likely to become confused.

Trademarks, which generally are renewable for as long as their owners wish to retain them, help consumers to identify products known to be of a desired quality and thus enable producers to profit from their products' reputations and brand names. A lot of money is spent creating and nurturing brand names. Because advertising has been considered one of the best ways to build and support brand equity, it's not surprising that well-known brands have big ad budgets. For example, in 1994 Philip Morris spent $907 million on brand advertising; AT&T spent $673 million; and Procter and Gamble spent nearly $516 million.

A measure of the value of brands by *Financial World Magazine* finds Marlboro ranking first and valued at $44.6 billion. Coca-Cola is second and worth $43.4 billion. McDonalds is third with a brand worth of $18.9 billion.

In an age of abundant products and information overload, the best brands stand out and reveal their value by creating consumer loyalty—repeat buyers of a company's products or services. Brand names, logos, packaging, and designs are elements of communications devised by companies to create awareness and acceptance of their products in the marketplace. Companies seek to protect these communications elements with trade or service marks.

Legal protection of trademarks against misuse and infringement is rooted in commerce, specifically in the common law tradition of unfair competition, and with today's state and federal statutes protecting trademarks and service marks and trade dress.

The Federal Trademark Statute

The Lanham Trademark Act of 1946 (15 U.S.C., Sect. 1051-1127, 1988 ed. and Supp. V) gives a seller or producer the exclusive right to register a trademark and prevent competitors from using that trademark. Registration is in effect for 10 years; an affidavit of continued use must be filed in the sixth year. Renewal can occur any number of successive 10-year terms so long as the mark is still in commercial use. The trademark must be used in interstate commerce, although federal protection still may apply if the trademark has an effect on interstate commerce.

A producer may apply for a trademark if there is an "intent to use" the mark; the use must occur in six months or an extension can be requested.

The Lanham Act says that trademarks "include any word, name, symbol, or device, or any combination thereof." A symbol or device may be almost anything that can carry meaning. Previously courts have ruled on and allowed the following as trademarks:

- A particular shape—a Coca-Cola bottle;
- A particular sound—NBC's three chimes;
- A particular scent—sewing thread with plumeria blossoms.

A phone number with a totally made-up name and backed by promotional efforts may qualify for a trademark.

A trademark may be refused registration if it

1. Is immoral, deceptive, or scandalous or would "disparage or falsely suggest a connection with persons, living or dead, institutions, beliefs, or national symbols, or bring them into contempt, or disrepute."

2. Uses the U.S. flag or other insignia or that of a state, city, or foreign country.

3. Uses the name, portrait, or signature that identifies a particular living person without his or her written consent or that of a president of the United States.

4. Is a mark "which so resembles a mark registered in the Patent and Trademark Office, or a mark or trade name previously used in the U.S. by another and not abandoned, as to be likely, when used on or in connection with the goods of the applicant, to cause confusion, or to cause mistake, or to deceive."

5. Is a mark that is merely descriptive of the goods or deceptively misdescriptive of them or primarily is geographically descriptive of the goods.

A trademark requires that a person or company use or intend to use the mark "to identify and distinguish his or her goods, including a unique product, from those manufactured or sold by others and to indicate the source of the goods, even if that source is unknown" (15 U.S.C., Sec. 1127).

Trademark protection exists at both the state and federal levels. About half the states have statutory or common law that cover misuse or dilution of trademarks. Again, for trademarks used in interstate commerce, federal protection is available under the Lanham Act.

The "use it or lose it" rule applies to trademarks and the mark owner is responsible for policing efforts. Failure to take action or ceasing to use the mark could be considered an abandonment.

In addition, the mark always must be placed on the product packaging and product literature. And the appropriate symbols, ® or ™, should appear next to the trademark or service mark.

Generic Names

A name becomes generic and loses trademark status when the primary significance of the term to the consuming public is the product rather than the producer.

Some names that have become generic include toll house (cookies), shredded wheat, (air) shuttle, thermos, aspirin, and Monopoly.

Color Meets the Legal Test for a Trademark

In 1995, the U.S. Supreme Court in *Qualitex Co.* v. *Jacobson Products Co., Inc.*, ruled that a color may be registered as a trademark under the Lanham Act. The court found that, sometimes, a color will meet ordinary legal trademark requirements. And, when it does, no special legal rule prevents color alone from serving as a trademark.

Qualitex had used (since the 1950s) a special shade of green-gold color on the pads that it made and sold to dry cleaning firms for use on dry cleaning presses. Jacobson Products, a competitor, began to sell its own press pad in 1989 using a similar green-gold. In 1991, Qualitex registered the special color with the Patent and Trademark Office and then filed a lawsuit against Jacobson for trademark infringement.

The Ninth Circuit Court ruled that the Lanham Act did not permit registration of color alone as a trademark. The Supreme Court disagreed, stating that color alone could meet the legal requirements for use as a trademark, because the language of Lanham Act "describes the universe of things that can qualify as a trademark in the broadest of terms . . . and by the underlying principles of trademark law, including the requirements that the mark identify and distinguish the seller's goods . . . from those manufactured or sold by others and to indicate [their] source . . . and that it not be functional."

Secondary Meanings

Color, like words or a scent, can attain a "secondary meaning" in the minds of the public and therefore identify and distinguish a particular brand. A functional product feature, however, cannot serve as a trademark "if it is essential to the use or purpose of the article or if it affects the cost or quality of the article."

A trademark, such as a particular color, "can act as a symbol that distinguishes a firm's goods and identifies their source, without serving any other significant function." This follows a tenet of trademark law that seeks to promote competition by protecting a firm's reputation, rather than a functional product feature. The latter would be the province of patent law. In the Qualitex case, the firm's green-gold color acts as a symbol, identifying to customers the press pads' source; the color serves no other function.

Trade Dress

Until recently, trade dress protection basically meant original packaging; now it refers to a product's total image. Competitors therefore can be stopped from imitating the general appearance or image of a product or service if the trade dress was either inherently distinctive or had acquired a secondary meaning.

In a 1992 case, *Two Pesos, Inc.* v. *Taco Cabana, Inc.*, the U.S. Supreme Court held that "trade dress which is inherently distinctive is protected under the Lanham Act without a showing that it has acquired a secondary meaning, since such trade dress itself is capable of identifying products or services as coming from a specific source."

In this case, Taco Cabana sought to prevent a competing restaurant chain from imitating its trade dress described as "a festive eating atmosphere having interior dining and patio areas decorated with artifacts, bright colors, paintings and murals . . . the stepped exterior of the building is a festive and vivid color scheme using top border paint and neon stripes. Bright awnings and umbrellas continue the theme."

Two Pesos adopted a motif that was very similar to the preceding description. Taco Cabana sued Two Pesos, saying that the infringement created a likelihood of confusion on the part of ordinary customers as to the source or association of the restaurant's goods or services.

Proving Infringement

Liability under Section 43(a) of the Lanham Act requires proof of the likelihood of confusion of the ordinary consumer about the product's source.

In the Qualitex case, the plaintiff submitted survey evidence showing that customers identified its press pad by the gold-green color. In the survey, 39 percent of the respondents identified Qualitex as the maker of the gold-green press pad.

The survey claimed to show that the press pad's color helped establish a secondary meaning that would qualify for trademark protection. The Supreme Court, as mentioned earlier, agreed to this and that rival Jacobson's product was so similar that it caused confusion to ordinary customers.

Taco Cabana in its suit against Two Pesos offered no survey evidence. The Supreme Court agreed with the court of appeals, reasoning that "while the necessarily imperfect (and often prohibitively difficult) methods for assessing secondary meaning address the empirical question of current consumer association, the legal recognition of an inherently distinctive trademark or trade dress acknowledges the owner's legitimate proprietary interest in its unique and valuable information device, regardless of whether substantial consumer association yet bestows the additional empirical protection of secondary meaning."

Crackdown on Counterfeiters

The Anticounterfeiting Consumer Protection Act of 1996 (PL 104-153), passed on July 2, 1996, seeks to control and prevent commercial trademark counterfeiting.

The software industry claims that sales of pirated software account for more than 40 percent of total revenue. The significance, scope, and consequences of counterfeiting include not only counterfeit goods; public health risks, money laundering, and organized crime also are involved.

The act improves the ability of law enforcement officers to detect and arrest counterfeiters. It allows for the meaningful prosecution of all levels of a criminal organization involved in counterfeiting.

The act increases criminal penalties by making trafficking in counterfeit goods or services a RICO (Racketeering and Corrupt Organization) predicate offense, placing counterfeiting activities within the RICO statute, thereby providing for increased jail time, criminal fines, and asset forfeiture.

The act involves all federal law enforcement and gives more authority to seize counterfeit goods and manufacturing equipment. Also, goods will be less likely to reenter the economy. The seized counterfeit goods will be destroyed rather than returned to the importer for reshipment to another port of entry.

The act provides for judicially determined statutory damages for trademark owners. Businesses that suffer commercial damages from counterfeit products may be awarded either actual or statutory damages. The act adds strength to existing statutes and provides stronger civil fines pegged to the value of genuine goods and statutory damage awards of up to $1 million per counterfeited trademark.

Congress concluded that the counterfeiting of trademarked and copyrighted merchandise:

1. Has been connected with organized crime;
2. Deprives legitimate trademark and copyright owners of substantial revenues and consumer goodwill;
3. Poses health and safety threats to United States consumers;
4. Eliminates U.S. jobs; and
5. Is a multibillion-dollar drain on the U.S. economy.

Section 3 of the act amends RICO Section 1961(1)(B) of title 18 U.S.C., by inserting a new section 2318 relating to trafficking in counterfeit labels for "phonorecords, computer programs or computer program documentation or packaging . . . " and section 2320 of the copy-right act relating to trafficking in goods and services bearing counterfeit marks, also sections 2314 and 2315 relating to interstate transportation of stolen property.

Section 7 of the act amends 15 U.S.C. Sect. 1117 to provide statutory damages as an alternative to actual damages. A plaintiff may elect, and a court may approve, statutory damages ranging from $500 to $100,000 per mark for each type of merchandise involved or up to $1 million per mark for each type of merchandise if the violation is willful.

Sections 8 and 9 of the act would forbid the Customs Service from reexporting piratical merchandise. Counterfeit goods bearing an American trademark would have to be destroyed. The act authorizes the Customs Service to impose a civil fine on persons who are in any way involved (direct, assist financially, aid and abet) in the importation of counterfeit goods for sale or public distribution. A first offense calls for a fine equal to the market value of the merchandise, if the merchandise were genuine. A second offense and thereafter calls for doubling the fine. Fines would be at the discretion of the Customs Service and in addition to any other civil or criminal penalty or other remedy authorized by law.

Trademark Dilution

The Federal Trademark Dilution Act of 1995 (PL 104-98) is aimed at protecting famous marks and allows injunctive action against a party that causes dilution of the distinctive quality of the mark. Dilution is defined as "the lessening of the capacity of a mark to identify and distinguish goods or services, regardless of the presence or absence of—(1) competition between the owner of the famous mark and other parties, or (2) likelihood of confusion, mistake, or deception."

Other legal action can be taken if it can be proven that the "person against whom the injunction is sought willfully intended to trade on the owner's reputation or to cause dilution of the famous mark." It is up to the court to determine damages, which could include triple damages.

Civil Actions and Sanctions for Trademark Violations

An injunction is designed to protect the property of the plaintiff and could be issued to restrain ongoing or future infringing acts. In addition to an injunction, the court may grant an order for the seizure of goods and counterfeit marks and the means of making such marks and the destruction of all materials and equipment.

In addition to any damages sustained by the plaintiff, the court may order the defendant's profits be given the plaintiff as well that the defendant pay the plaintiff's legal costs.

An injunction can be issued against an organization or its agent, officers, or employees.

Trademark Protection Under GATT

In Article 15 of GATT, *trademarks* are defined as "any sign or a combination of signs capable of distinguishing the goods or services of one undertaking and those of another."

Article 16 provides the exclusive right to prevent all third parties from using, in the course of trade, identical or similarly confusing marks.

The initial registration, and its renewals, shall be for a term of no less than seven years with renewals afforded in perpetuity (Sect. 2 Art. 18).

Under GATT, trademark registration can be canceled for nonuse only after an uninterrupted period of at least three years. Under NAFTA, the nonuse period is two years.

Service marks are given the same protection as trademarks. The TRIPS agreement, however, does not include NAFTA's specific references to "design, letters, numerals, colors, figurative elements, or the shape of goods or their packaging" (Article 1708, 1).

Compulsory licensing of trademarks shall not be permitted (Sect. 2 Art. 21), and owners of registered marks are free to assign their marks with or without a transfer of the business to which the mark applies.

Trademark Protection Strategies

The basic method for protecting a trademark is to continuously use the mark.

Next, monitor third-party use. A trademark search usually is used when a company is trying to avoid copying a name or mark that is in use and registered with the Patent and Trademark Office. This proactive approach can also be used to monitor possible infringement of your mark.

With almost a million currently registered or pending marks, trademark search software and databases can make the job easier. Because a trademark may be a word, phrase, symbol, slogan, character, shape, or combination, you need to select the category of your mark as well as a class of goods or services. You can search for either an exact match or for similarly sounding marks.

TRADEMARKSCAN, an on-line database, lists pending state and federal trademark applications and registrations. Data from the U.S. Patent and Trademark Office, states, and other countries are now available on-line and on CD-ROM. MarkSearch from MicroPatent and Trademark Access from CCH Trademark Research are CD-ROM programs used for searches.

Other sources are the U.S. Patent and Trademark Office, on-line indexes such as NEXIS, and directories such as the Thomas Register.

All of the aforementioned should be considered preliminary trademark search sources, as there are common law trademarks that are in use but not registered.

Failure to undertake a comprehensive trademark search before marketing a product or service could be very expensive, should a registered or common law mark later turn up. Then, of course, there could be an infringement problem and court-imposed sanctions for failure to make a good faith effort to uncover a competing mark.

U.S. Patent Law

A patent protects any new or useful process, machine, manufacture or composition of matter, or any new and useful improvement. A computer program is a process or an integral part of a machine, and is therefore patentable.

Computer hardware usually meets the substantive requirements of the Patent Code (see U.S.C., Title 35—Patents). Patent law covers not only the code but interpretations of that code by the federal courts and the regulations and actions of the Patent Office. The term of a patent is 17 years from the date of issuance.

Patents Under the General Agreement on Tariffs and Trade (GATT)

Section 5 of the GATT agreement provides patent protection for "any inventions, whether products or processes, in all fields of technology, provided that they are new, involve an innovative step and are capable of industrial application" (Sect. 5, Art. 27.1).

The term of patent protection shall be at least 20 years (Sect. 5, Art. 33) counted from the filing date. Under current U.S. law, patent protection runs 17 years from the date of patent issue. Objections to the 20 years from filing date clause center on the length of time it sometimes takes the government to approve a patent—which can be 10 years or more.

Article 27 requires that "patents shall be available and patent rights enjoyable without discrimination as to the place of invention." This implies a proof of a date of invention for establishing priority of invention.

Semiconductor Layout Designs

Article 36 of the GATT agreement gives recognition of the protection of layout designs and topographies of integrated circuits. It is unlawful if, within a member's territory, "a person imports, sells, or otherwise distributes for commercial purposes: a protected layout-design, an integrated circuit in which a protected layout design is incorporated, or an article incorporating such an integrated circuit."

Article 36 requires authorization of the right holder to sell or import a protected topography or an integrated circuit with a protected layout or design. Article 38 provides a 10-year minimum period of protection but allows for protection for up to 15 years. Also, there are limitations on the granting of compulsory licenses for semiconductor technology.

Industrial Designs

Protection is given to "independently created industrial designs that are new or original" (GATT, Article 25, Section 4).

6

Copyrights

The Copyright Act of 1976 is a revision of the Copyright Act of 1909; U.S.C. Title 17 became fully effective on January 1, 1978.

In 1980, copyright legislation extended protection to computer programs (17 U.S.C. 107, 117). Computer programs are copyrightable as "literary works."

To come under copyright protection, a work must be

1. In the form of a "writing," that is, fixed in any tangible medium of expression now known or later developed, from which it can be perceived, reproduced or otherwise communicated, either directly or with the aid of a machine or device;

2. A product of original creative authorship.

Computer programs and software fall under copyrightable works. The definition of *literary works* (Section 101 of the act) refers to works expressed in "words, numbers, or other verbal or numerical symbols or indicia." A computer program is defined in Section 101 as "a set of statements or instructions used directly or indirectly in a computer in order to bring about a certain result." It should be noted that only the programmer's "literary" expression, that is, the program, would be copyrightable, not any "procedure, process, system, method of operation, concept, principle, or discovery, regardless of the form in which it is described, explained, illustrated, or embodied" (Sect. 102(b)).

Copyright grants the owner the exclusive right to do and to authorize others to

- Reproduce copies of the copyrighted work;
- Prepare derivative works based on the copyrighted work;
- Distribute copies of the copyrighted work to the public by sale or other transfer of ownership or by rental, lease, or lending;
- Perform the copyrighted work publicly; and
- Display the copyrighted work publicly (17 U.S.C. 106).

Infringement of Copyright and Criminal Acts

It is an infringement of the copyright for anyone to engage in any of the activities just listed without the authorization of the copyright owner.

Under copyright law, there are four kinds of criminal acts:

1. Infringement of "a copyright willfully and for purposes of commercial advantage or private financial gain";

2. Intentional fraudulent use of a copyright notice, whereby copyright notice is placed on an article when the defendant knows the notice to be false;

3. Fraudulent removal of copyright notice;

4. Knowingly making a false representation in a copyright registration application.

Criminal copyright can also arise from the distribution of infringing goods.

In a criminal prosecution for copyright infringement, the government must prove that a copyright was infringed, that the violation was willful, and that the infringement was for profit.

Criminal and Civil Penalties

The 1992 amendments to the copyright act stiffened criminal penalties: convicted first offenders may get prison sentences for up to 5 years; fines of up to $250,000 for individuals, $500,000 for an organization; with a previous conviction, the maximum prison sentence could be 10 years.

A forfeiture clause allows the courts to seize and destroy infringing items plus "all implements, devices or equipment used in the manufacture of . . . infringing copies."

Court Decisions on Fair Use and Preemption

Recent court decisions have examined two important doctrines in the federal copyright law: fair use and the federal preemption of state statutes that protect copyrightable information. The preemption doctrine has critical implications for state computer crime statutes and for organizations charging unlawful activity under these statutes. The threat is that charges could be dismissed if the state law provides criminal sanctions for copyright violations.

Before discussing these problems in detail, we give a brief review of the relevant sections on fair use and preemption.

Fair Use Doctrine

Our copyright system is based on the dual interests of property rights and intellectual promotion. Fair use makes the copyright law flexible rather than a rigid doctrine and a law that does not impede or stifle creative activity.

Congress codified the doctrine of fair use in the Copyright Act of 1976; however, it did not define the term, leaving its interpretation to the courts. Section 107 of the copyright act gives four factors that courts may consider in deciding whether a particular use is fair:

1. The purpose and character of the use, including whether such use is of a commercial nature or is for nonprofit educational purposes;

2. The nature of the copyrighted work;

3. The amount and substantiality of the portion used in relation to the copyrighted work as a whole; and

4. The effect of the use upon the potential market for or the value of the copyrighted work.

Congress has expressed that these four factors are neither all-inclusive nor determinative but can provide "some gauge for balancing equities." These factors, therefore, are a flexible set of guidelines for the courts to use in analyzing and deciding each individual copyright infringement case where the issue is one of fair use.

Courts must evaluate whether the use of copyrighted material was of a commercial nature or for a nonprofit educational, scientific, or historical purpose; whether the nature of the copyrighted work was published in a tangible form or was unpublished material; what amount, in quantity and substantiality—its core or essence—of the work was used; and whether the defendant's alleged conduct had an adverse effect on the potential market or value of the copyrighted work.

Cases Test Fair Use Doctrine in Copyright Law

In 1992, the U.S. District Court for the Southern District of New York heard a class action suit brought on behalf of six publishers against Texaco Inc. for copyright infringement. The publishers claimed a Texaco research scientist had infringed copyrights by making photocopies of journal articles for his personal work files. The sole issue before the court was the fair use doctrine in Sections 107 and 108 of the copyright act.

The district court looked at the purpose and character of the use and ruled against Texaco's argument that the use was "productive" and advanced scientific discovery. The court found the copying to be "superseding in nature," not productive.

Examining the factor of market effect of the photocopying, the court also ruled against Texaco. That Texaco did not get a license to photocopy journals from the Copyright Clearance Center, and pay royalties for in excess of fair-use copies, may have influenced the court's decision.

On an appeal to the Second U.S. Court of Appeals, the broad issue of whether photocopying of scientific articles is fair use was rejected by the court. Instead, the court confined its ruling to the narrower issue of "archival" or long-term use photocopying, and in a 2-to-1 ruling affirmed the district court's decision (*American Geophysical Union* v. *Texaco Inc.*).

The decision by the appeals court does not widen the copyright protection for publishers as to the photocopying of scientific articles nor did the courts touch on the copying practices of nonprofit entities.

Copying and Distribution via Computer

In-house copying and computer distribution of newsletters and other published materials violates the copyright law. Electronically storing copyrighted material in a company database and sharing it with personnel via e-mail could easily jeopardize the entire network (see the *Phillips Business Information Inc.* v. *Atlas Telecom Inc.* case). A company found guilty of criminal copyright infringement—copying and distribution—could find "all . . . equipment used in the manufacture of . . . infringing copies" seized and destroyed by order of the court.

Be aware that some publishers offer bounties to employees or other insiders who provide details on illegal copying.

The Preemption Clause

The first Copyright Act of 1790 gave the United States a dual system of copyright: federal law protected published works; common law or state law protected the body of unpublished works. The Copyright Act of 1909 preserved the common law right of authors in their unpublished works. The Copyright Act of 1976 ended this dual system by attaching federal copyright protection to a work the moment it is fixed in tangible form. The federal government also was given complete responsibility for enforcing copyright law.

Congress could have looked to the Supremacy Clause of the Constitution (Article VI, clause 2) to end the dual system of federal and state law but instead chose to legislate explicitly on preemption.

The key provision on preemption is Section 301 of the act of 1976, which ends copyright protections under the common law or statutes of any state.

In the House Report on the 1976 copyright bill, the intention of section 301 is "to preempt and abolish any rights under the common law or statutes of a state that are equivalent to copyright and extend to works coming within the scope of the Federal copyright law. The declaration of this principle in section 301 is intended to be stated in the clearest and most unequivocal language possible, so as to foreclose any conceivable misinterpretation of its unqualified intention that Congress shall act preemptively, and to avoid the development of any vague borderline areas between State and Federal protection."

Title 28 U.S.C., Section 1338, makes it clear that any action involving rights under the federal copyright law come within the exclusive jurisdiction of the federal courts.

For a state law to be preempted, two conditions must be met:

1. The state right must be "equivalent to any of the exclusive rights within the general scope of copyright as specified by section 106."

2. The right must be "in works of authorship that are fixed in a tangible medium of expression and come within the subject matter of copyright as specified by sections 102 and 103" (discussed previously in the text).

Two general areas are left unaffected by the preemption:

1. Subject matter that does not come within the subject matter of copyright;
2. Violations of rights that are not equivalent to any of the exclusive rights under copyright.

The House Report gives examples of actions and rights that are not preempted and different from copyright, including misappropriation, as long as it "is in fact based neither on a right within the general scope of copyright as specified by section 106 nor on a right equivalent thereto." The 1976 act, in effect, "preserved rights under state law with respect to activities violating rights that are not equivalent to any of the exclusive rights within the general scope of copyright." These include misappropriation, trespass, conversion, breaches of contract or trust, defamation, invasion of privacy, deceptive trade practices, breaching the security of a computer database, and electronic interception of data.

States may not add to their criminal statutes an additional element that creates a distinguishable offense from the proscriptions in the copyright act.

Complete Preemption and Computer Crime Laws

In 1993, in *Rosciszewski* v. *Arete Associates Inc.*, the Fourth U.S. Circuit Court of Appeals held that the protection of computer programs from unauthorized copying under the Virginia Computer Crimes Act is completely preempted by the federal copyright act.

Rosciszewski had filed an action in a Virginia state court and under the Virginia act alleging that Arete personnel had broken Rosciszewski's computer security system and had stolen copies of a copyrighted software program. Arete got the action removed to federal district court in Virginia on grounds that the action was a federal copyright matter. Rosciszewski moved to have it remanded it back to state court.

Specifically, the Virginia act made it a crime to gain access to a computer network without authority and with the intent to obtain property or services by false pretenses or to convert the property of another.

The Fourth Circuit Court determined that the core of the complaint was the unauthorized copying of a computer program. In essence, the court said that the complaint did not require proof of elements beyond those necessary to prove that the copyrighted program was infringed. The court held that the criminal intent element of the Virginia act "does not add an element qualitatively changing the state claim from one of unauthorized copying." The act's intent element, therefore, did not add an element creating a distinguishable offense.

A state cause of action can be preempted under the conditions described earlier—that the work at issue is copyright material and that state law created an equivalent right within the general scope of copyright.

The Fourth Circuit Court found that the removal of the case from the Virginia state court to a federal court was proper.

Comment: State computer crime laws may be in violation of the U.S. Constitution and preempted if property is defined to include copyrighted information or criminal acts that involve copyright material. Cases in state courts could be dismissed.

The alternative, of course, is to file possible criminal copyright infringements in federal courts under the copyright act. As discussed earlier, the federal criminal and civil penalties are harsh. But, before bringing any case to a federal attorney, you must know precisely what has been infringed, have specific allegations, and even some solid evidence. The feds do not have a lot of resources—training, personnel, expertise—in intellectual property law violations or prosecutions. Be prepared to do your own investigation. Examine your state's computer crime law and look at other statutes covering theft, conversion, and conspiracy.

On-Line Liabilities: Copyright and Trademark Infringements

In November 1995, the United States District Court for Northern California ruled that Netcom On-Line Services and a bulletin board service (BBS), Usenet, should have taken action to prevent a Usenet subscriber, Dennis Erlich, from continuing to commit copyright infringement after Netcom and Usenet had been told to do so by the copyright owners.

Netcom, Usenet, and Erlich were sued for copyright infringement by the copyright holders, Religious Technology Center and Bridge Publications, Inc. The works were published and unpublished Church of Scientology materials.

Usenet is a BBS that uses Netcom to gain access to the Internet. Netcom claims it neither creates nor controls information content and does not monitor posted messages. Netcom claimed Erlich made the copies of the copyrighted material and posted them. Netcom, therefore, moved for summary judgment.

Summary judgment is appropriate under the Federal Rules of Civil Procedure (56(c)) "if the pleadings, depositions, answers to interrogatories and admissions on file, together with the affidavits, if any, show that there is no genuine issue as to any material fact and that the moving party is entitled to judgment as a matter of law." The moving party has the burden of showing that no genuine issues of material fact are present that should be decided at trial.

Netcom's contention that it did not directly infringe copyrights and therefore should not be held liable was agreed to by U.S. District Judge Ronald M. Whyte. He said Netcom was similar to the owner of a copy machine, which should be examined for contributory, not direct infringement: "there should still be some element of volition or causation which is lacking where a defendant's system is merely used to create a copy by a third party."

However, both Netcom and Usenet failed to take any action against Erlich after they were notified of the copyright infringements. The court ruled that Erlich's postings far exceeded any "fair use" of the copyrighted works.

Netcom could be held liable for contributory infringement, said Judge Whyte, if the plaintiffs could prove Netcom knew of the infringement and failed to stop it, because its failure to "cancel Erlich's infringing message and thereby stop an infringing copy from being distributed worldwide constitutes substantial participation in Erlich's public distribution of the message."

This issue of material fact would have to be resolved at trial. (Ultimately, it was settled out of court.)

Another issue is whether Netcom materially contributed to the infringement or was vicariously liable. Judge Whyte concluded that Netcom did not control the infringer's acts and did not derive a financial benefit from the infringement. This eliminated a finding of vicarious liability.

The extent of liability issue to be decided at trial hinges on Netcom's "knowing" about copyright infringement and how it should have reacted.

There seems to be general agreement that an access provider such as Netcom cannot screen or monitor everything that comes on to the system, but it should have an obligation to respond and act when it is notified of intellectual property violations.

Bulletin Board Service Held Liable for Copyright and Trademark Infringements

In *Playboy Enterprises, Inc.* v. *Frena* (1993), the defendant, Frena, operated a subscription computer bulletin board service, charged a monthly fee, and allowed subscribers to upload and download material from files on the board. A subscriber to Frena's service uploaded 170 photographs that were copyrighted by Playboy Enterprises, Inc. (PEI). The photos were displayed on Frena's BBS and downloaded by some of his customers.

Frena denied that he ever uploaded any of PEI's photographs and that, as soon as he was made aware (via a summons) that the photos were copyrighted, he removed the photos from his BBS.

To establish copyright infringement, PEI had to show ownership of the copyright and "copying" by Frena. Ownership was easily shown as well as Frena's access to the photos. Frena supplied the product, his BBS, that contained the unauthorized copies of the photographs and publicly distributed and displayed the photos. The concept of display covers "the projection of an image on a screen or other surface by any method, the transmission of an image by electronic or other means, and the showing of an image on a CRT, or similar apparatus connected with any sort of information storage and retrieval system."

A "public display" is a display open to the public, persons outside the normal circle of family and friends. Frena's display of the copyrighted material was to his subscribers.

In Frena's case, the fair use argument was rejected because the BBS was a for-profit venture, a commercial use that could result in future market harm to PEI.

Trademark Infringement on BBS

In this case, Frena also infringed on PEI's registered trademarks by using "Playboy" and "Playmate" in file descriptors for the photos. Frena claimed this was done by a subscriber. Frena did admit to removing PEI text on the photos and substituting his company's name and phone number.

The court found that PEI's trademarks met all the requirements: that it was distinctive, well-known, and strong. Frena claimed he did not intend to use PEI's trademarks. The court rejected this because a showing of intent or bad faith is not necessary to establish a trademark violation.

Unfair Competition Charge: "Reverse Passing Off"

By obliterating PEI's trademarks from the photos, Frena made it appear as though PEI had authorized Frena's product. This false inference and description, in effect, is a denial of PEI's right to public credit for the success and quality of its products. Such a "reverse passing off" is a violation of Section 43(a) of the Lanham Act, which allows federal action for unfair competition.

In awarding partial summary judgments to PEI, the court ruled that Frena had infringed on PEI's copyrights and trademarks and had violated the Lanham Act.

Copyright Infringement: Individual and Organizational Liability

Organizations today are faced with lessened standards of criminal liability and with courts that increasingly find a connection between parties—corporations, directors, officers, employees, and agents—laying negligence on any or all of these for failing to be aware of risks or to protect their customers or vendors from harmful acts. The key phrase here is *aware of risks*. The organization and its management should be aware of the risks to its customers or suppliers inherent in business operations.

Unauthorized copying of vendor software can lead to charges of criminal copyright infringement. When organizations purchase software, they obtain a license agreement that grants them a nonexclusive right to use the software. Copying software beyond the bounds of the license agreement could find the organization guilty of breach of contract or copyright infringement. The penalties can be severe and legal costs exorbitant. Litigation also can adversely affect the public perception of a company, damage credit lines, and harm employee, customer, and investor relations.

The sections that follow deal with criminal copyright infringement and the potential liability of organizations, directors and officers, and managers for not having software copying policies, for inadequate monitoring of licensed vendor software, as well as for failing to promote among employees recognition that unapproved copying of vendor software is unethical and illegal.

Directors and Officers Duties and Liability

The basic legal duties of corporate directors are loyalty and care; that is, first, avoiding conflicts of interest and, second, being informed about company operations and not making poorly considered decisions or being negligent.

The relationship of the corporate director and officer to stockholders is similar to that of an agent to a principal. And, the liability involved is similar, in that it may be based on failure to perform a statutory or common law duty. Failure to use ordinary care and prudence, when it results in loss, can generate liability.

Standards of Liability

General standards of liability and burdens of proof include these:

Strict liability—This involves the act only; it requires no proof of intent to commit the act. Copyright infringement is a strict liability matter; innocent infringement is still infringement. A director or manager can be held personally liable for willful or knowing infringement if he or she took part in the copying.

Vicarious liability—In general, corporations are vicariously liable for the actionable conduct of their employees performed in the scope of their employment. This traditional doctrine applies to aiding and abetting a crime or a conspiracy to commit a crime and acts with the knowledge and intention of facilitating the commission of a crime. Copyright infringement involves a financial gain. *Contributory infringement* means one knowingly induced or contributed to the act.

Derivative liability—The acts and intent of corporate officers and agents are imputable to the corporate entity.

Willful blindness or indifference—Here the corporation or its officers and agents intentionally avoid knowing a situation or act that will incriminate. Willfulness is a disregard for the governing statute and an indifference to its requirements.

Flagrant organizational indifference—This is the conscious avoidance by an organization of learning about and observing the requirements of a statute.

Rogue employee—This is someone who, for his or her own benefit, commits an illegal act or whose conduct violates company policy and procedure, despite in-place efforts to prevent such an act.

Responsible corporate officer doctrine—This doctrine applies to any corporate officer or employee "standing in responsible relation" to a forbidden act. Liability can arise if the officer could have prevented or corrected a violation and failed to do so. Strict liability, for the act only, applies; no mental element is involved.

This is a critical doctrine with significant implications: an officer has a positive duty to seek out and remedy violations when they occur and a duty to implement measures that will ensure that violations will not occur. The doctrine forces the corporate officer to define which risks he or she should know, because this person is likely to be held to an affirmative duty of care concerning those risks.

Risk Awareness

How does one foresee a risk? It is hard to foresee a very specific risk. How can a director or officer know that software copying is a widespread problem, in specific industries or in computer-dependent companies? What matters in law is that those in responsible positions within an organization be aware of the risks that might affect vendors or customers and that appropriate precautions are taken to alleviate those risks.

This, in short, is a definition of the *duty of care*. But what should be the level of "knowing"? (Or, what should a reasonable person have known?) The answer is that it is the director's or manager's duty to find out if copying of vendor software is a problem in the company. Not seeking to know could lead to liability. It is expected that directors, officers, and managers will monitor employee conduct for violations of law.

Liability also can arise from awareness of the imminent probability of specific harm to another. This is legally referred to as *notice*, or having specific knowledge concerning the existence of a fact or condition. Notice may give rise to a duty to protect or at least to investigate the situation.

Liability for Inadequate Protection

Organizations can be held liable for inadequate protection of copyright material if there is evidence that

1. The infringement was similar to a previous violation committed by the company and its employees (this could mean a failure to supervise).
2. The organization did not take all economically feasible steps to provide a reasonable level of protection (copying policies, monitoring, and audits).

Lessening an Organization's Liability Risk

An organization's basic legal defense strategy against criminal copyright infringement should include written software copying policies and codes of conduct that discourage and deter unethical and illegal behavior. The policies and codes should be distributed to management and employees and contain specific prohibitions. Such policies and codes can reinforce legal norms and have a positive effect in deterring unlawful behavior in the organization. Codes must be enforced. Enforcement procedures also should be spelled out so that violations get reported, investigated, and disposed of.

Because criminal copyright infringement comes under the U.S. Sentencing Commission Guidelines (see Sect. 2B5.3), a wise choice for organizations is to follow the Sentencing Commission guidelines for compliance programs as described in Chapter 8 of the *Federal Sentencing Guidelines Manual*.

Sample Software Copying Policy Content

This policy of the corporation is to cover the entire and internal corporate licensed software usage, including microcomputers, terminals, and networks. The sole purpose of the corporation's use of licensed software is to assist in conducting the business of the enterprise. Licensed software is to be considered "business property" and is to be used for business purposes only.

All computers and communications equipment and facilities and the data and information stored on them are and remain at all times business property of the corporation and are to be used for business purposes only.

The corporation devises and maintains the security of its computing and communications systems as well as the monitoring of such systems, including licensed software. The use of all licensed software must be made known to the corporation.

The corporation reserves the right to assign usage of licensed software in any manner and to monitor all software usage.

Violation of corporate policies by employees will invoke disciplinary measures up to and including termination of employment. This policy will be reviewed periodically and updated in light of new legal developments and corporate experiences.

Promulgation of the Software Copying Policy

1. Disseminate a copy of the policy to all those whose conduct it is to govern.
2. Write it clearly, so it is easily understood.
3. Have each recipient sign off on it, with initials and the date.
4. It can be communicated via
   a. An employee handbook—maintain a log of employees who receive the handbook;
   b. Posting on employee bulletin boards;
   c. The physical distribution of the printed text.

Each part of the software copying policy should be explained in detail:

- Define the nature of the license and contract and why it is important to the organization.
- Explain the copyright law and the possible sanctions employees, officers, and the organization could face for infringement or breach of contract.
- Desirable and undesirable behavior should be explained, described, and examples of each given.
- The annual software copying compliance audit should be described and reasons given for its necessity. Also explain any ancillary benefits of the audit, such as possible cost savings.
- Give the reasoning behind and describe unannounced audits.
- Explain the need for centralized control of software purchases, installations, and use monitoring.
- Describe how an employee can report unauthorized software copying and retain confidentiality.
- Detail the sanctions for unauthorized software copying.

Monitoring and Auditing

The organization must take reasonable steps via monitoring and auditing systems that will detect criminal conduct by its employees and agents. It should use annual and unannounced software copying audits that would detect and deter possible infringement as well as informal mechanisms related to organizational structure and management controls, including centralized administration of software purchases and usage. In decentralized companies, branches and subsidiaries would need similar controls and audits.

Audits of the Policy

To monitor compliance with policy directives and procedures (the purpose of the audit),

1. Set the frequency and timing of audits;
2. Conduct them with an internal audit department;
3. Focus on formal and informal management controls and assess their effectiveness;
4. Thoroughly review all controls in each area;
5. Document and report the audit's findings.

Reporting Systems

The sentencing guidelines also require that the organization create and publicize a system for reporting criminal conduct within the organization that allows employees to do so without fear of retribution.

This requirement can be handled in several ways. A policy directive from corporate management should clarify when to report possible copyright infringement, under what circumstances, and to whom.

Enforcement

Disciplinary mechanisms must be established for violations of law as well as for the failure to detect an offense. The "form of discipline that will be appropriate will be case specific."

Violations of Policy

Enforcement and sanctions should

1. Be consistent in application;

2. Provide disciplinary mechanisms for illegal conduct, unethical conduct, and failure to detect an offense;

3. Define conduct that is grounds for termination of employment;

4. Ensure that disciplinary measures do not conflict with employment laws;

5. Ensure that termination actions do not conflict with the personnel policy manual (consult with legal counsel on termination actions).

This section assumes that everyone will be aware just what and when software copying is a violation of law.

Adequate discipline, obviously, can conflict with union rules and employment laws. Personnel policy manuals may have to state that illegal or unethical conduct is grounds for termination of employment. However, any manager should consult with legal counsel before firing an employee.

7

Information Privacy and Confidentiality

A1995 poll found that 83 percent of American adults are concerned about invasion of privacy; 53 percent are very concerned about such threats. Consumers who have a high degree of expectations for privacy cited concerns regarding financial relationships (72 percent) and health information (71 percent). Consumers indicated a demand for privacy protection measures: those who felt such measures were very important or somewhat important totaled 94 percent for health organizations and 93 percent for financial services companies.

Consumers most concerned about privacy have the following demographic characteristics: people with incomes above $35,000, college graduates, and those with postgraduate education.

The poll, conducted by Louis Harris Associates for Privacy and American Business, involved 2,506 consumers over the age of 18 and has a potential error factor of ±2 percent.

Another indication of current public feelings about privacy of personal information is the strong response to an alleged breach of privacy. In September 1996, a computer database called P-TRAK, operated by Lexis-Nexus to provide information to lawyers to use when locating heirs, litigants, and persons related to a case, raised a privacy issue furor on the Internet. Postings claimed the database was being used to identify an individual by their name, social security number, mother's maiden name, and other personal information. Lexis-Nexis said it was releasing only names, addresses, telephone numbers, and sometimes, maiden names.

The extent of the outrage over whether social security numbers and other sensitive personal information is being divulged for commercial purposes was a surprise to many privacy experts. One result of this incident is that three senators have requested a study "of possible violations of consumer privacy rights by companies that operate computer databases . . . The companies are allegedly obtaining and compiling personal background data on individual citizens into electronic-transmitted databases for sale to private entities or individuals, including attorneys, banks and credit card companies, without the consent of the targeted individuals."

The study should determine if "the non-consensual compilation, sale and usage of databases is a violation of private citizens' civil rights . . . and if there are ways consumers can prevent databased service companies from including their personal background information in commercial databases absent their consent."

The senators asked the Federal Trade Commission to report on whether there is a need for new legislation.

Privacy as a Right

The concept of a personal right to privacy can be traced to an article in the *Harvard Law Review* in 1890 by Louis D. Brandeis and Samuel D. Warren. They advanced the idea that enough case law existed for recognition of a formal legal right of privacy. Fifteen years later, the Georgia Supreme Court ruled that a man named Pavesich could recover damages from the insurance company that had used a picture of him in an advertisement and testimonial without his permission.

The development of a common law right to privacy and of a privacy invasion tort has been given in the Restatement (Second) of Torts, Sec. 652B (1977), as follows: "One who intentionally intrudes, physically or otherwise, upon the solitude or seclusion of another, or his personal affairs or concerns, is subject to liability to the other for invasion of his privacy, if the intrusion would be highly offensive to the ordinary reasonable person."

Disclosing personal and private information by a record keeper, depending on the circumstances, could lead to a suit for negligence.

Defamation

Wrongful disclosure of private or embarrassing facts usually requires such information to be communicated to more than one person. Any disclosure of false information could lead to a defamation suit. Defamation has two types of communication: defamation via writing, called *libel,* and *slander,* which is defamation by speech. Both are communication of false information to a third party that injures a person's reputation. Other elements of defamation include the reasonable identification of the defamed person and damage to the person's reputation; if the defamation refers to a public figure or is a matter of public concern, it must be proven that the defamatory language was false and that it was communicated knowingly or with a reckless disregard as to the truth or falsity of the information.

The Privacy Act of 1974

When Congress examined the issue of privacy in the early 1970s, there was a growing public awareness of actual and potential abuses of personal privacy, particularly in the area of information gathering and dissemination by public agencies and private companies, such as commercial credit bureaus. In 1974, Congress passed the first federal legislation providing privacy protection.

The Privacy Act of 1974 covered federal agencies and government contractors performing record keeping functions on behalf of a federal agency. Federal agencies were required to provide safeguards for an individual against an invasion of personal privacy. Unless exempted by law, federal agencies were required to

1. Permit an individual to determine what records about him or her are collected, maintained, used, or disseminated by such agencies;

2. Allow the individual to prevent such records from being used by other agencies without his or her consent;

3. Allow the individual access to his or her records, to copy, correct, and amend them;

4. Collect, maintain, use, or disseminate any record of identifiable personal information only for necessary or lawful purposes and ensure that the information is current and accurate for its intended use;

5. Be exempt from the act's requirements only where specific statutory authority exists;

6. Be subject to civil suit for violations of an individual's privacy rights and damages for willful or intentional acts.

Privacy Protection Policies and Practices

An effective privacy protection policy must have three concurrent objectives:

To minimize intrusiveness—It must create a proper balance between what an individual is expected to divulge to a record-keeping organization and what that person seeks in return. The organization needs to tell the individual its information needs and collection practices as well as how it will use reasonable care in selecting and retaining other organizations to collect information about individuals on its behalf and the organization's information controls and safeguards.

To maximize fairness—It must minimize the extent to which recorded information is itself a source of unfairness in any decision made about the individual on the basis of it. This means the person has a right of access to his or her records for reviewing, copying, and correcting.

To create a legally enforceable expectation of confidentiality—It must develop and define obligations as to the uses and disclosures that will be made of recorded information about an individual, restricting the record keeper's discretion to voluntarily disclose a record about an individual.

Since 1974, federal and state laws have incorporated these principles of privacy protection. Most laws cover the individual's right to see and copy information collected about him or her, to correct or amend such information, and to seek redress of grievances or injury caused as a consequence of the use of inaccurate data.

Record-keeping organizations always must be concerned with the accuracy of their information, that it is up-to-date and complete. Data collection and dissemination and data security are all-important. Organizations are responsible for verifying data it collects and correcting any false information it knowingly passes on to another party—organizations cannot argue a general presumption of accuracy regarding third-party data they use or transmit.

The collection and dissemination of all information generated during the employment and postemployment periods should be scrutinized in terms of the broad privacy protection policy just discussed.

Eight Principles of Privacy and Information Protection

1. Information should not be collected unless its need and relevancy have been clearly established.

2. Information should not be collected fraudulently or unfairly.

3. Information should be used only if it is accurate and current.

4. Individuals must know the existence of information stored about them, why it has been recorded and how it is collected, used, and disseminated; and they have the right to examine that information upon request.

5. There must be a clear procedure on how the individual can correct, delete, or amend inaccurate, obsolete, or irrelevant information.

6. An organization collecting, maintaining, using, or disseminating personal information should assure its reliability, integrity, and availability and take precautions to prevent its misuse.

7. There must be a clear procedure and safeguards to prevent personal information collected for one purpose from being used for another purpose or disclosed to a third party without an individual's consent. There also must be a right to notification of disclosure of information.

8. Only legally authorized personal information should be collected by federal, state, and local governments.

Employee Privacy and Security

The following are areas of contention between security and individual privacy:

1. Arrest and criminal records;
2. Bank information (checking and savings accounts, safe deposit boxes, loans);
3. Credit information reporting;
4. Medical records;
5. Employment records;
6. Tax records;
7. Education records;
8. Use of investigative agencies or background investigations;
9. Use of a polygraph (at any stage of employment);
10. Use of psychological stress evaluator (PSE);

11. Use of psychological tests;
12. Use of a voice print analyzer;
13. Use of personal identification or verification methods;
14. Use of "service monitoring" methods or equipment;
15. Use of electronic surveillance equipment;
16. Secret or investigative files or dossiers;
17. Internal control, screening, or detection procedures or equipment.

An employee personnel file must contain job related information only, yet comply with various federal wage-hour and EEO record-keeping requirements as well as record retention policies set by existing federal, state and local laws or regulations.

Federal Laws Affecting Employee Privacy and Security

Many federal laws and regulations affect the employee, the workplace, and security measures. Note that the following summarize critical legislation and judicial rulings with which managers should be familiar:

- Most federal laws apply to employers with 15 or more employees.
- With antidiscrimination laws, discrimination does not have to be intentional for an employer, or its agent, to be in violation of the law.
- Every legislative act has been followed by many, and often conflicting, regulations and rulings of administrative agencies, and court decisions.
- Laws usually are not clarified for years; that is, it takes a build-up of case law before a law's full intent and substance are revealed.

The Age Discrimination in Employment Act (29 U.S.C. 621 et seq. (1967)) prohibits discrimination on the basis of age. Inquiries about age have been considered evidence of discrimination against applicants in protected age group—40–65 years of age. The act covers hiring, referral, classification, and other conditions of employment. Firms with 25 or more employees fall under the act, and the administrative agency is the Department of Labor.

The Cable Communications Policy Act of 1984 protects the privacy of cable TV subscriber records. In the Cable Act of 1992, the privacy protections of 1984 act are extended to cable companies that may provide cellular and other wireless services. Companies are required to inform subscribers of the nature of personally identifiable customer information collected and the nature of the use made of that information. Third-party disclosure, with some exceptions, requires customer consent.

The Communications Assistance for Law Enforcement Act (CALEA) (PL 103-414; 47 U.S.C. 1001-1010) will facilitate law enforcement agency surveillance in digital, wireless, cellular, and other advanced communications technologies and services. The law requires telecommunications carriers to assist the lawful interception needs of law enforcement within four years, plus a two-year extension for cause. CALEA requires common carriers and telecommunications manufacturers to provide the necessary equipment to meet the technical assistance needs of the FBI and other law enforcement agencies to carry out electronic surveillance. This assistance is to be accomplished while protecting the privacy of communications and without impeding the introduction of new technologies and services.

CALEA requires telecommunications carriers to ensure that their systems can isolate expeditiously the content of targeted communications and information, identifying its origin and destination; provide this information to law enforcement agencies so it can be retransmitted; and carry out intercepts unobtrusively, so targets are not made aware of the interception, and in a manner that does not violate the privacy and security of other communications.

The Computer Security Act of 1987 provides for the security and privacy of sensitive information in federal computer systems through the development of security standards, research, and training, and the establishment of computer security plans by all systems operators. The act also provides for a Privacy Advisory Board.

The Computer Matching and Privacy Protection Act of 1988 targets computer matching programs conducted by federal agencies in which computerized information from two databases is automatically compared to detect common individual data or records or discrepancies in records. The act established agency data integrity boards to review, approve and monitor matching programs.

The Debt Collection Act of 1982 places restrictions on the release of federal debt information to private credit bureaus.

The Driver's Privacy Protection Act of 1994 prohibits the use of state motor vehicle lists by parties other than government agencies, law enforcement agencies, and the judicial system. Statistical market researchers may have access and use lists "so long as the personal information is not published, redisclosed, or used to contact individuals." State motor vehicle agencies "shall not knowingly disclose or make available to any person or entity personal information about any individual obtained by the department in connection with a motor vehicle record."

The act requires an opt-out opportunity for drivers, meaning states must offer a way for drivers to indicate they do not want their names on mailing lists.

Under the act, an individual may sue a state motor vehicle department for giving out personal information and the firm that uses such information without specific permission.

The Electronic Communications Privacy Act of 1986 extends the protection of the Wiretap Act of 1968 to electronic communications, both stored and transmitted. The privacy of electronic mail is protected; for providers of electronic communications services to the public it is illegal to divulge knowingly the contents of any communication except to the sender or intended recipient of the information.

The Employee Polygraph Protection Act of 1988 (29 U.S.C. 20001-20008) prohibits the use of lie detectors or any similar device "for the purpose of rendering a diagnostic opinion regarding the honesty of an individual." "Lie detectors" include polygraphs, deceptographs, voice stress analyzers, psychological stress evaluators, and "any similar device for the purpose of rendering a diagnostic opinion regarding the honesty or dishonesty of an individual."

It is unlawful to

1. Make any employee or prospective employee take a lie detector test;

2. Use or get the results of any lie detector test of an employee;

3. Discharge or take disciplinary action or discriminate against any current or prospective employee who refuses to take a lie detector test, or who fails a lie detector test, or who files a complaint under this law.

The law lays out the rights of the examinee, how the test should be conducted, the qualifications and requirements of examiners, and the disclosure of test results.

Violations of this law carry a civil penalty of not more than $10,000. However, the employer is open to civil actions by the employee and may be liable for back pay and benefits, court costs, and the employee's attorney's fees. The statute of limitations is three years after the date of the alleged violation.

Among the exemptions under the law are these:

- Government employees;
- Defense Department contractors and consultants;
- Intelligence or counterintelligence personnel, consultants, or contractors;
- Limited exemption is given to private security services engaged in armored car transportation, alarm security systems installation, utilities security, public transportation security, transportation of currency and precious commodities, and controlled substance manufacturers or distributors;
- Firms and agencies involved in ongoing investigations, providing the employee is thought to be involved in an economic crime and the employer details in a statement to the examinee the reasons for the test.

To be regarded as "ongoing," the investigation must deal with a specific incident and an "economic loss or injury" must have occurred. An investigation, therefore, cannot be part of a sustained surveillance program, and the loss must have happened and be documented via report, audit, or initial incident investigation. The economic loss could result from theft, embezzlement, fraud, or industrial espionage or sabotage.

Employees who had access to the property under investigation and of whom the employer has a "reasonable suspicion" may be subject to lie detector or polygraph testing.

The individual conducting the polygraph testing must have a valid, current state license and be bonded.

Note again, through all of this, that the burden is on the employer to show that all the various elements exist to conduct polygraph tests. Therefore, employers are urged to consult with legal counsel prior to using lie detection testing.

The Equal Employment Opportunity Commission (EEOC), which administers the Equal Employment Opportunity Act (42 U.S.C. 2000 et seq. (1972)), issued guidelines that relate to any selection techniques that may be improperly used so as to have the effect of discriminating illegally against racial, ethnic, religious, age, or sex groups. The EEOC guidelines require that any employment selection procedure be supported by validity and by evidence of "a high degree of utility."

The Family Educational Rights and Privacy Act of 1974 limits disclosure of educational records from schools and colleges that get federal funds.

The Fair Credit Reporting Act (FCRA) (15 U.S.C. 1681 et seq. (1970)) regulates the methods of obtaining credit information on an applicant or employee. Essentially, the law requires that an applicant must be informed in writing that a credit report is being sought and the information sought must be defined.

Under FCRA, a credit bureau may furnish an employer a report for "employment purposes," defined as a report used for the purpose of evaluation only. Section 615 of FCRA requires the report user, if it denies employment to an individual based wholly or partly on information in the report, to advise the individual and give him or her the name of the credit agency that made the report.

A bill to shore up the savings and loan insurance fund was signed into law (PL 104-208) on September 30, 1996. A provision modifies the 1970 Fair Credit Reporting Act, which governs credit bureaus and the compilation of individuals' credit histories.

The law gives consumers new and more definite ways of correcting errors in credit reports. Consumers now can get a free credit report after their credit history has been used for an action that would deny them credit for any class of transaction, such as insurance or employment. The information that caused adverse action, upon written request from the consumer, must be disclosed to the consumer within 60 days. Credit bureaus must delete disputed information if it cannot be verified.

The law now covers banks, retailers, and other businesses that supply credit bureaus with consumer information. These businesses are required to take steps to ensure the accuracy of the information collected and investigate disputed information. Liability can be avoided if the person or company can show "by a preponderance of the evidence that at the time of the alleged violation he maintained reasonable procedures to assure compliance" with the act's accuracy and privacy safeguards.

Affiliated companies can share consumer information for marketing purposes.

There is a requirement that the Federal Reserve Board study whether the release of a person's Social Security number and other identifying information increases the chances for fraud against financial institutions.

The Federal Contract Compliance Regulations (E.O. 11246 as amended by Executive Order 11375 (1968)) prohibit discrimination on the basis of race, color, religion, sex, or national origin and mandate employers take "affirmative action" to ensure equal opportunity provisions.

This executive order applies to companies with federal contracts or subcontracts for more than $10,000; companies with over $50,000 in federal contracts and 50 or more employees are required to file affirmative action plans with the Office of Federal Contract Compliance.

The National Labor Relations Act (29 U.S.C. 151 et seq.) covers unfair labor practices. The act established the National Labor Relations Board, which investigates and arbitrates charges of unfair labor practices, generally in the area of union organizing and an employer's attempts either to thwart union organizing attempts or to inquire of or influence an employee's actions regarding unions.

The Occupational Safety and Health Act (OSHA) (29 U.S.C. 651 et seq. (1970)) originally was set up to protect workers from unsafe working conditions. Recently, however, OSHA has sought, through court action and rules, to allow workers and the agency broad access to employer records relating to workers' safety and health.

The Privacy Act of 1974 restricts federal agencies from releasing personal information and allows an individual review of files.

The Right to Financial Privacy Act of 1978 limits the access of law enforcement agencies to an individual's bank and financial records.

The Telephone Consumer Protection Act of 1991 restricts the activities of telemarketers.

Title VII of the Civil Rights Act of 1964 defines discrimination in comprehensive terms as any employment practice that has an adverse impact on the members of any protected class. It should be noted that discrimination does not have to be intentional for the employer to be in violation of the law (see *Griggs* v. *Duke Power Co.*, 401 U.S. 424 (1971), 3 EDP, Section 8137). The Equal Employment Opportunity Commission (EEOC) was established as the administrative agency for Title VII.

The Video Privacy Protection Act of 1988 protects privacy of video rental records.

Legislation regarding wiretap access, added to the 1994 Crime Bill, requires telephone company networks to meet the wiretapping needs of law enforcement agencies. Law enforcement groups cannot obtain e-mail addresses by subpoena; and they have to get a court order to learn the caller location of a cellular or wireless user.

Security and personnel managers should examine the provisions of the following federal laws, which have an impact on the employment process in terms of discrimination, privacy, or record keeping:

The Americans With Disabilities Act of 1992, which says that any health information obtained by employers must be stored in separate files and treated as confidential;

The Employee Retirement Income Security Act (ERISA) (P.L. 93-406, 88 Stat. 829 (1974));

The Equal Pay Act of 1963 (and the amendment to the Fair Labor Standards Act of 1938);

The Consumer Credit Protection Act (15 U.S.C. 1601 et seq. (1968));

The Labor-Management Reporting and Disclosure Act (29 U.S.C. 401 et seq. (1959));

The Rehabilitation Act (29 U.S.C. 701 et seq. (1973));

The Vietnam Era Veteran's Readjustment Assistance Act (41 U.S.C. 60-250, 38 U.S.C. 2012)).

New Health Care Law Has Security and Privacy Mandates

The Health Insurance Portability and Accountability Act of 1996 (PL 104-191) was signed into law on August 21, 1996. The purpose of Subtitle F of the act is to improve "the efficiency and effectiveness of the health care system by encouraging the development of a health information system through the establishment of standards and requirements for the electronic transmission of certain health information." The act also calls for standards to protect the privacy of personal health information.

In the act, individually identifiable health information means any information collected from an individual that relates to the past, present, or future physical or mental health or condition, as well as health care and payment provisions, and identifies the individual, or provides enough information from which reasonable basis for identification can be made.

Information Standards

Standards are to be promulgated that reference a data element of health information or a transaction. Adopted standards will apply to a health plan, a health care clearinghouse, and a health care provider who transmits health information in electronic form.

The standards apply to financial and administrative data elements and transactions. A standard is to be set for a unique health identifier for each individual, employer, health plan, and health care provider and the purposes for which the identifier may be used.

Security standards, before adoption, must consider the costs of security measures, the extent of training needed, the value of audit trails in computerized record systems, the needs and capabilities of small and rural health care providers, and ensuring that a health care clearing-house has adequate security policies and procedures.

Under the act, any person or organization that maintains or trans-mits health information "shall maintain reasonable and appropriate ad-ministrative, technical, and physical safeguards" that will ensure the integrity and confidentiality of the information, protect against "any reasonably anticipated" threat or hazard to the security and integrity of the information and unauthorized uses or disclosures of the informa-tion, and ensure compliance by the officers and employees with safe-guards.

Penalties for Noncompliance and Violations

A person who knowingly violates the confidentiality of a unique health identifier, or obtains individually identifiable health information, or discloses such information to another person can be fined up to $50,000 or imprisoned for one year or both. If the offense is committed under false pretenses, the maximum fine rises to $100,00 and the possible prison sentence goes to five years. If the offense is committed with the intent to "sell, transfer, or use" such information for "commercial ad-vantage, personal gain, or malicious harm, the fine jumps to a maxi-mum of $250,000 and the prison sentence to not more than 10 years, or both.

Health Information Privacy Recommendations

The act requires the Secretary of Health and Human Services to come up with detailed privacy standards for individual health information. The recommended standards should address, at a minimum, the fol-lowing:

1. The rights of an individual to personally identifiable health infor-mation;
2. Procedures to exercise these rights;
3. The authorized uses and disclosures of such information.

The final standards are to be enacted by legislation within 36 months; if not, the Secretary must issue final regulations within 42 months after the date the act was enacted.

No federal laws cover access to medical records and only a few states allow patients the right to review and copy their medical records.

Thus, if a medical record is incorrect, the patient never has the opportunity to address those errors. Currently, the only protection of medical records is under a patchwork of state laws; 34 states have protection laws, with 28 state laws providing patients with access to their medical records.

However, legislation already has been introduced in the form of the Medical Records Confidentiality Act (S.1360). In addition, earlier proposals from the Clinton administration called for privacy standards covering "individually identifiable health information that is in the health information system."

Privacy principles to be incorporated in the standards on a person's health information include

1. Prohibitions against unauthorized disclosures, unless there is a program exception or by the enrollee or with the enrollee's permission, or for law enforcement purposes;

2. No more than minimal disclosure, only the amount "necessary to accomplish the purpose for which the information is being disclosed";

3. No disclosure of risk-related data, information that could be used to set premiums based on risk adjustment factors;

4. Provisions for security; users and holders of health information must "implement administrative, technical, and physical safeguards for the security of such information."

Individuals or enrollees in a health care program also must have the right to know who uses or maintains their health information and for what purposes. The enrollee would have the right to inspect, copy, and amend or correct personal information. And an enrollee or representative would have the right to get a written statement from the health care provider concerning the purposes for which such information may be used, disclosed, or accessed.

Personal enrollee health care information may not be used in making employment decisions.

8

Records Management

Records management deals with the organization, control, storage, and retrieval of business records for their life cycle, particularly for the length of time required by law.

Information systems departments must be aware of the legal requirements that go with the creation and storage of electronic documents and records. These legal issues have become acute because of the growing demand by government agencies for records that prove or disprove an organization's regulatory compliance. Therefore, records management must be integrated into and a key part of an organization's compliance program.

Previously, much of the responsibility for records retention fell to the records management department. It organized and stored records for the length of time required by law and established safeguards against the loss of records. Even though some laws do not specifically refer to computer-generated records, it has become an accepted business practice to store relevant information on computers. Record managers are familiar with federal requirements for archiving electronic materials.

Most business records are either vital for business necessity and recovery or for corporate memory. The new problem is how to handle documents and records that could affect compliance.

Vital records have three essential elements: the information content, the processing activity, and the output function, the activity that results from processing—a billing mailed, for example. Vital records also are those records that, if destroyed, would seriously impair the company's ability to resume operations quickly.

In a data processing disaster, normally the loss of information or the inability to use that information when needed is most critical. Hardware losses are important, of course, but equipment often can be replaced easier and faster than information can be recovered.

This chapter presents an overview of the various laws and regulations that require enterprises to keep records, to take precautionary measures to avoid loss of important records, and to make speedy recovery from a disaster involving information systems. It also examines the circumstances where litigation can pose a serious risk to the organization having a records management program that is less than comprehensive and efficient.

The two vital elements of a records management program are, first, control of information along with adequate information storage, retrieval, and archiving and, second, speedy recovery of information and data processing operations. Record retrieval can become urgent when serious issues arise, such as during an audit or a discovery request in litigation. In litigation, specific records, regardless of the storage media, may have to meet evidence foundational requirements. For electronic records, this normally means the accuracy, reliability, and trustworthiness of the system that created the record.

Several of the laws that will be cited in this chapter do not specifically refer to computer systems or to computer-generated data and records. However, most organizations process and store relevant business information on computers; therefore, current interpretations of the legal term *records and accounts* have included optical and magnetic computer recording media.

Laws and regulations applicable to data storage and recovery seek to define the scope of information that a business must protect in order to produce certain records upon legitimate demand and those records that a business must have to continue operating as a going concern.

If the business cannot produce its records or continue to operate because records have been destroyed, then it is likely that questions of negligence and liability will be raised.

In the first part of this chapter, we examine federal statutes requiring record keeping and computer disaster preparedness. In addition, we look at several laws that imply the need for record keeping, as well as a new e-mail records retention standard.

Record Keeping for Government Compliance

Over 1,000 federal statutes and regulations require the retention of records and documents. The major portion of document and media retention programs are based on requirements that are reasonably spelled out in various laws. Retention requirements can range from one week to permanently; most retention requirements, though, are for three to five years.

All laws and regulations covering an organization's activities should be examined to determine specific types of records to retain and maintain.

The Foreign Corrupt Practices Act (FCPA)

The FCPA (Title I, PL 95-213, 91 Stat. 1494) of 1977 amended the Securities and Exchange Act of 1934 by imposing on companies registered with the SEC certain record-keeping and internal control standards, originally designed to prevent companies from hiding foreign bribery payments. The "accounting standards" provision of the act requires affected companies to

1. Make and keep books, records, and accounts that, in reasonable detail, accurately and fairly reflect the transactions and disposition of the assets of the issuer; and

2. Devise and maintain a system of internal accounting controls sufficient to provide reasonable assurance that (a) transactions are executed in accordance with management's general or specific authorizations, (b) transactions are recorded as necessary to permit preparation of financial statements in conformity with generally accepted accounting principals or any other criteria applicable to such statements, (c) access to assets is permitted only in accordance with management's general or specific authorization, and (d) the recorded accountability for assets is compared with the existing assets at reasonable intervals and appropriate action is taken with respect to any difference.

In subsequent releases to the act, the SEC said that the FCPA was intended to deal with a much broader range of practices than the problem of illegal corporate payments. The FCPA is the major piece of federal legislation that spells out the requirement for keeping books and records.

In 1988, the Omnibus Trade and Competitiveness Act amended the FCPA in several areas including standards for what is permissible conduct under the act. Adequate documentation and record keeping of transactions are vital elements of the act; the question of due diligence would be applied if a company's records were destroyed in a disaster.

SEC Regulations

The Securities and Exchange Commission requires investment companies to keep and maintain accurate records of stock transactions and to safeguard the required records. Requiring records to be promptly available and secure implies having disaster prevention measures, including off-site records storage, a disaster recovery plan, and possibly a data processing backup site. The basic requirements are spelled out in 17 C.F.R. Section 270.31a-2, 275.402-2 and subsequent sections.

Office of the Comptroller of the Currency (OCC)

The OCC has several bulletins on data processing disaster recovery for banks and financial institutions. The OCC has issued recommendations that banks have alternate data processing capability, including having off-site backup areas, disaster recovery plans of their own, or making sure that their service bureaus have contingency plans. The OCC also requires that a bank's recovery plan be testable and that it be reviewed on a yearly basis.

The banking overhaul law requires larger banks and thrifts to undergo annual independent audits of their financial statements and compliance with laws and regulations affecting accounting standards, statements, and records.

Other agencies that regulate financial institutions, such as the Federal Home Loan Bank Board, have requirements that limit the amount of time a bank has to either return a check or produce a record, which implies the necessity for having plans and procedures for secure record keeping and disaster recovery.

Uniform Commercial Code

The Uniform Commercial Code (UCC) governs commercial transactions, and every state has enacted the code in whole or in part. One part of the UCC limits the time a bank has to return a dishonored check; the time limit can be exceeded if the bank's operations are disrupted by a disaster or other circumstances beyond its control. The bank, however, can be excused only "provided it exercises such diligence as the circumstances require." This provision of UCC 4-108 has been used to infer that banks should have disaster recovery plans to meet the necessary standard of care to avoid liability.

Under UCC 4A, which covers commercial wire transfers, the duty to keep records of transactions is implied in the standard. Although deliberately written loosely, UCC 4A, both by the process involved in a computer or communications funds transfer and by the security procedures mandated, requires that records of transactions be protected and archived. Interestingly, these requirements of due diligence would seem to fall on both the customer and his or her bank and any intermediary bank.

The Federal Reserve and the Treasury Department

The Bank Secrecy Act of 1970 (12 U.S.C.S., Sect. 1730d, 1829b, and 1951–1959) requires financial institutions to maintain records of the identity of a customer and microfilm checks on accounts of over $100, as well as other records.

The Federal Deposit Insurance Act (12 U.S.C.S., Sect. 1829b) requires insured banks to keep "appropriate records and procedures." 12 U.S.C.S., Section 1953(a) requires any uninsured bank or institution to keep "records or evidence of any type . . . that have a high degree of usefulness in criminal, tax, or regulatory investigations or proceedings." A record-keeping violation that is willful or grossly negligent can draw a civil penalty of up to $10,000. The criminal penalty for a violation can be a fine of $1,000 or one year in prison or both. In 1995, the Federal Reserve and the Treasury Department jointly adopted record-keeping requirements related to certain wire transfers by financial institutions; the rules took effect April 1, 1996.

Under the Bank Secrecy Act and the amendments of the Annunzio-Wylie Anti-Money Laundering Act of 1992, the Treasury Department and the Federal Reserve are authorized to require financial institutions to maintain records regarding domestic and international funds transfers. Such records are to be retained for five years. Wire transfers for less than $3,000 are exempted.

The originator's bank must retain, for each payment order accepted, the originator's name and address, the amount, date, payment instructions received with the payment order, beneficiary bank identification, and if available, the beneficiary's name and address or account number.

For other than established customers, banks are to verify the name and address of the originator, the verification document's number (driver's license, social security card, passport, employer, taxpayer, or alien identification).

In addition to storage, banks must have access to and be able to retrieve records "within a reasonable period of time," depending on the nature of the request and the quantity and type of record.

Another Federal Reserve regulation calls for an Automated Clearing House (ACH) to have adequate contingency backup capabilities "to cover equipment failure or other developments affecting the adequacy of the service being provided." This includes record retention and availability of records. Also, an adequate audit review of the program must be done at least annually.

The Money Laundering Suppression Act of 1994 expanded reporting coverage to checks, drafts, notes, and money orders drawn on a foreign financial institution. Currency transaction reports are required for businesses "engaged in providing check cashing, currency exchange, or money transmitting or remittance services, or issuing or redeeming money orders, travelers' checks, and similar instruments . . . " And, casinos were made financial institutions for reporting purposes.

The Electronic Funds Transfer Act (EFTA)

This federal statute (92 Stat. 3728, 15 U.S.C., Sect. 1693 et seq.) covers a wide range of electronic funds transfers, including point-of-sale transactions and other consumer payments. Under the statute, banks must take reasonable precautions against acts that could disrupt the operation of the transfer or affect the records related to such transactions.

ERISA

Under the Employee Retirement Income Security Act (ERISA) employers can set up a tax-deferred, income plan known as the 401(k). This plan allows companies to give employees responsibility for their own investing decisions. A new section of ERISA, 404(c), is an optional regulation. Administration of these plans may be done by the company in-house or by an outside plan administrative service provider. Administering 401(k) plans is an intensely computer-driven process. Administrators may have to provide on-line information regarding an employee's account balances on investments and savings and investment choices and holdings, initiate fund transaction decisions, and modify the investment portfolio.

The 404(c) regulations have a section on confidentiality relevant to data processing operations. Information relating to employer securities investment purchases, holdings, sales, voting rights, and other aspects of a employee's portfolio must be maintained "with procedures designed to safeguard the confidentiality of such information." Section

404(c) requires that a fiduciary be designated who will be responsible for ensuring that the safeguards are being followed. The employee also must receive a description of the safeguard procedures along with the name and address of the plan fiduciary responsible for monitoring compliance with the procedures.

Tax Records

Section 6001 of the IRS Code requires income taxpayers to keep records to show they have complied with tax laws.

The IRS requires that "all machine-sensible records whose contents may be or may become material to the administration of the Code shall be retained by the taxpayer. The retained records shall be in a retrieval format that provides the information necessary to determine the correct tax liability."

The taxpayer is responsible for ensuring that all source documents underlying the accounting data be identified and made available to the IRS on request. Documentation also must be kept on the computer system and materials that will help the IRS understand or operate the system, if necessary. The IRS may also review internal controls and security measures "associated with the creation and storage of the records."

All computer-generated records must be clearly labeled and stored in a secure environment. Backup copies should be stored at an off-site location.

Organizations that use EDI must retain computer records "that, in combination with any other records, contain all of the detailed information required by section 6001." This information may be captured at any level within the accounting system "provided the audit trail, authenticity, and integrity of the retained records can be established."

Contract-Related Records

Government contractors face a host of regulations that require retention and recoverability of records that are computer generated and telecommunicated. These records now include contract information, bids, specifications, engineering drawings, billings, and payments.

Americans With Disabilities Act (ADA)

Under the ADA, any health information obtained by employers must be stored in separate files and treated as confidential. Section 102(B) says "information obtained regarding the medical condition or history of the applicant is collected and maintained on separate forms and in separate medical files and is treated as a confidential medical record."

No mention is made of how long these records must be kept, but they must be made available on request to "government officials investigating compliance with this Act." This is a critical clause in many administrative laws.

E-Mail Records Management Standard

The National Archives Records Administration (NARA) has issued proposed standards for management of federal records created or received on e-mail systems. The standards cover all federal government agencies on the proper means of identifying, maintaining, and disposing of e-mail records. Although covering federal agencies, the standards should be examined by private companies for guidance in setting up records management programs.

The standards were prompted by the 1993 court rulings in *Armstrong* v. *Executive Office of the President*, where the federal archivist was ordered to provide guidance and assistance to federal agencies on preserving e-mail records.

Definitions of a *Document,* an *Electronic Record,* and an *E-Mail Message*

The Federal Records Act establishes two conditions that must be met for a document to be a record: (1) the document must be made or received by agency personnel under federal law or in connection with the transaction of public business, and (2) it must be preserved or appropriate for preservation. NARA defines an *electronic record* as "numeric, graphic, text, and any other information recorded on any medium that can be read by using a computer and satisfies the definition of a Federal record in 44 U.S.C. 3301." An *e-mail message* is "a document created or received on an e-mail system including brief notes, more formal or substantive narrative documents, and any attachments, such as word processing documents, which may be transmitted with the message."

Not all e-mail messages will meet the statutory definition of records. NARA cautions, however, that "e-mail messages are not considered nonrecord materials merely because the information they contain may also be available elsewhere on paper or in electronic files . . . in addition, multiple copies of records may all be records if they are used for different purposes in the conduct of official business or filed in different files."

In addition to the text of messages, e-mail systems often provide transmission, receipt, and acknowledgment data as well as sender and the date the message was sent and any distribution list or directory; this also must be preserved. Agencies are encouraged to incorporate these features into their e-mail systems.

Draft Documents

Drafts must be maintained for adequate and proper documentation if "they contain unique information, such as annotations and comments, that helps explain the formulation or execution of agency policies, decisions, actions, or responsibilities; and they were circulated or made available to employees other than the creator for the purpose of approval, comment, action, recommendation, follow-up, or to keep staff informed about agency business."

Calendars and Other Materials

E-mail systems that contain calendars, indexes of events, and task lists for users also may be federal records. If these materials relate to high-level officials, NARA will appraise their value for future use.

Messages on Network Services

Use of an e-mail system on the Internet or other commercial service by a government agency "does not alter in any way the agency's obligation under the Federal Records Act." It is up to the agency to ensure that such records are preserved.

E-Mail Record Keeping

Agencies must maintain all e-mail records in an appropriate record-keeping system that will allow easy and timely retrieval, easily separate record and nonrecord material, and retain the record in a usable format until it should be disposed of. There must also be a records backup system.

Various media can be used for storage, electronic or paper. NARA, however, will determine which format is best for the preservation of records.

E-mail records may not be deleted or otherwise disposed of without prior disposition authority from NARA. This means all versions of e-mail records, including the original and all copies. Records may be destroyed only through the issuance of the General Records Schedules or the approval of schedules developed by agencies for records unique to the agency.

E-mail system users must be instructed on the required steps to be taken to ensure that the screen record is forwarded to the record-keeping feature of the system. If the system lacks this feature, the record must be forwarded to another record-keeping system. After the "live" screen record has been forwarded and stored, it can be deleted from the user's e-mail system.

Security of E-Mail Records

NARA calls for adequate security for records in e-mail systems, measures to protect e-mail records from unauthorized alterations or deletions, and regular backups for messages stored on-line.

Employee Training

Agencies must ensure that all employees are familiar with the legal requirements for the creation, maintenance, and disposition of e-mail records; know the agency's specific record-keeping requirements; and be able to distinguish between records and nonrecords on the systems.

Audits of how users identify records and maintain internal records management should be conducted by individual agencies. Audit reports should be made available to NARA on request and when it conducts evaluations of the agency's records management program.

Litigation Risks in Records Destruction and Retention

This brings us to an area where document retention and destruction policies may become involved in possible litigation or criminal proceedings involving the organization.

We take personnel materials as an example of the complexity of record-keeping issues where the potential for litigation is high. Obviously, these issues are not limited to the personnel area; similar problems can be found in those departments of the organization that deal with regulations and laws affecting product development, safety, environment, and antitrust.

The ADA and other discrimination laws, for example, have generated many lawsuits against employers. The ADA is enforced under the Civil Rights Act of 1964. This means the same rights and remedies under the Civil Rights Act apply to ADA. The responsible agencies are the EEOC and the Justice Department. Sections 709 and 710 of the Civil Rights Act detail procedures for conducting investigations and inspection of records.

Destruction of discoverable evidence—intentionally, recklessly, or through negligence—can bring charges of spoliation. Although a litigant is under no duty to retain every document, after a complaint has been filed, there is a duty to preserve what is known, or reasonably known, to be relevant or discoverable.

Spoliation has been charged under the obstruction of justice statutes, such as 181 U.S.C. 1510, which proscribes "preventing the communication . . . of information relating to a violation of criminal law." Destroying discoverable evidence can also lead to court-imposed sanctions, including

- tort remedies;
- specific discovery sanctions such as contempt of court, default judgments, or impositions of costs;
- instructions to juries to draw unfavorable inference against the destroyer of evidence.

Confidentiality and Discovery Problems

Recent court decisions have led personnel specialists to conclude that the best strategy in a discrimination suit is to have everything documented; that is, whether with an applicant or employee, every exchange should be recorded starting with the first contact. This means that, for instance, in a grievance situation, all information from initial complaint to hearing should be documented. With potential employee hires, documentation should start with the initial interview or screening test through explanations of the employee manual and the like. Only by keeping detailed records of every contact can the employer hope to present its side of the case.

The problems for computer protection and records management are threefold:

1. Notes and draft material need to be preserved and protected both for the privacy of the employee and for the confidentiality of the staff and company preparing a report.

2. When the report is finished, both notes and reports should be protected and stored in a safe place.

3. If a lawsuit is filed against the employer, both the report and draft material could be subject to a discovery motion by the plaintiff.

The last item often poses a dilemma from the perspective of information control. On the one hand, the information could be vital in defense of the company, yet some material could show the company in a bad light. And, potentially, every document created on the employee or applicant may have to be given to the employee's or applicant's attorney. But destroying discoverable evidence could lead to obstruction of justice charges or court-imposed sanctions.

Conclusion

Understanding the legal implications of producing and storing electronic documents has become mandatory for information services managers and systems users. Regulatory compliance, civil and criminal laws, and discovery requirements and sanctions are issues ignored at a firm's peril.

In the face of possible litigation, the best policy is to have documents reviewed by legal counsel.

The duty of computer protection managers is to

- Advise the personnel staff of the importance of computer security when creating notes or reports, stressing both the privacy and confidentiality aspects;

- Advise the personnel staff on the proper way to store computer media;
- Consult with the records storage staff on the organization's document management policies.

Although documents may be destroyed for legitimate reasons, the law is unsettled with regard to appropriate and inappropriate situations. A general rule is that a litigant should retain documents known or believed to be relevant or discoverable. However, a comprehensive and regular document *destruction* program should be in place.

Organizations must know and follow the mandated legal and regulatory requirements for records retention. But a key part of any records management program should be the regular, periodic destruction of records that are not crucial to business decisions or financial health; such "documents" often can be found on e-mail, in incident reports, and in memos, and they could pose a legal risk at some unknown point in time. These should be considered as candidates for destruction as shortly after birth as possible.

Creating Trustworthy Records

Checklist for Record Keeping

	YES	NO
1. The system, physical or electronic, must capture a record's		
Content—what is in the record;	☐	☐
Structure—the mechanism, machine, hardware, and software used to create the record;	☐	☐
Context—the business or legal activity for which the record was created and its purpose.	☐	☐
2. The process or system producing records must be shown to be reliable and accurate; specifically that a record was produced:		
As part of a regularly conducted business activity, at regular times, or as a regular activity but at times that were irregular;	☐	☐
By methods that ensure accuracy or where accuracy could be demonstrated via audits, regular monitoring, or quality control;	☐	☐
With timeliness, that is, simultaneously or within a short time after the activity;	☐	☐

By system procedures that accurately reveal the steps
in creating, modifying, duplicating, retention, and
destruction.   □ □
3. All procedures, controls, monitoring, audits, and training
programs associated with records management should be
documented.   □ □

Evaluating Off-Site Storage Facilities

Off-site storage facilities offer a variety of sites, protection and services.
In selecting a facility you must first know your requirements for records
retention, including

- How much storage space is needed?
- How must records be cataloged?
- How often will they be sent or brought to the facility?
- How must they be stored?

Once you have your requirements known, you can do a more effective evaluation of the facility.

Generally, an off-site facility should be at a distance from your
company, not to suffer a possible same disaster, yet close enough to retrieve important records quickly. Is the storage facility open to its clients
24 hours a day, 365 days a year?

There obviously are different levels of security. Security for an off-
site storage facility can be evaluated by site, access control, intrusion detection, fire suppression, and storage.

Site security can consist of fencing, perimeter lighting, CCTV, out-
door intrusion detection plus guards at gates. Does the facility have
these or can you simply walk up to the door?

The basic idea of *access control* is to identify with as much certainty
as possible the person seeking entrance to the facility.

- What access control system does the facility use?
- If it is card based, what type of card is used, those with magnetic
stripes or something more secure?
- Does the system log all attempts by individuals seeking access?
- How fast can access privileges be changed?
- How fast can new cards be issued?

Is the interior of the storage facility protected by an *intrusion detection* system and closed circuit television that is monitored?

- How is the system monitored?
- Is the system monitored 24 hours per day, seven days per week?
- Is the intrusion detection system wired directly to the police department?
- If it is in-house and proprietary, what is the guard response force?
- If the system is tied into a central station service, whose is it and what is its U.L. rating? Is the central station operated 24 hours a day?

What type of *fire detection and fire suppression* system is installed?

- Again, who monitors the system, and is it for 24 hours a day?
- Is the fire detection system wired to the fire department?

What equipment provides environmental controls: for temperature, humidity, air conditioning, dust control, and water or humidity detection?

Are the vaults, safes, or other *storage* systems rated for burglary, fire, or media? Remember, a "fire-proof" safe can cook its contents.

Media control handles each unit of media individually. Is there periodic inventory of stored media? Is there an effective media maintenance program?

Retrieval and transport systems may be needed at any time. Is there 24-hour per day accessibility, 7 days per week? Is there 24-hour per day courier service, 7 days per week?

- Are transport vehicles
  Environmentally controlled?
  Equipped with intrusion detection and alarms?
  Equipped with a fire suppression system?
- What is the maximum response time?

A facility should offer services designed to make the task of *records management* easier. Inventory of records, retention periods, removal authorizations, and maintaining record pickup schedules are basic services. Does the facility offer clients on-line access to the complete inventory of their records? Can the facility's software be integrated with the client's in-house document management program?

Some facilities offer copying of optical and magnetic media and have shredders, pulpers, and disintegrators for destroying obsolete records or degaussing for erasing magnetic media. Are clients given an affidavit certifying the destruction of records?

9

Disaster Planning for Information Systems
Legal and Regulatory Aspects

Most organizations would probably suffer a critical or total loss of function within two weeks if they lost their computer support. Failure to develop and implement contingency planning for data processing operations can have negative legal consequences for a company or its directors and officers. Management should recognize the essential procedures of a disaster recovery plan:

- Identification of critical information;
- Recovery of critical information;
- Designation of alternate sites;
- Plan testing and evaluation.

Management also should be aware of the various sources of possible legal liability. First is the basic fiduciary duty to protect the assets of the enterprise. Second, there are federal statutes, including the Foreign Corrupt Practices Act (FCPA), that have requirements for record keeping and internal controls.

Legal sanctions could arise from a situation where there was a loss of records and data processing operations. For example, a company is hit by a natural disaster that causes its data processing operation to shut down and also destroys information processed and stored in the computer. The shutdown is severe and long enough to result in a loss of business of such extent that survivability of the company is threatened.

In this example, the data processing performance breakdown could be seen as evidence of failure to "make and keep books, records, and accounts" or as weak or nonexistent internal controls. The corporation's stockholders, faced with a substantial drop in their stock price, would probably see that the situation resulted from negligence on the part of corporate officers and sue.

Even though the FCPA does not mandate specific internal controls, company controls will be judged by whether or not they are reasonable under the circumstances for the operation of business. In other words, was it reasonable under the circumstances for our hypothetical company not to have a contingency plan? Theoretically, liability could befall the company if factual circumstances lead to the conclusion that, because the company was computer dependent (assuming much of its accounts processing and record keeping was computerized), it should have had contingency planning to assure quick recovery of data processing operations after a disaster.

With the FCPA, the burden of compliance lies in understanding the term *reasonable* and the actions, or lack thereof, that follow from this. In business practice, then, what is judged "reasonable"? Generally, a demonstration evidencing reasonableness under the business judgment rule would be the presence of regular, competent, and periodic cost-benefit analyses of alternative contingency plans. Management also should be prepared to demonstrate that the contingency plan was implemented, tested, reviewed, and if necessary, modified. An audit review attesting to this would strengthen management's claims.

The Comptroller of the Currency's requirement that all national banks have a contingency plan is one of several legal obligations to ensure that banks continue to operate. Other applicable laws are contained in the Uniform Commercial Code and the Federal Electronic Funds Transfer Act.

The Employee Retirement Income Security Act of 1974 (ERISA) places statutory liabilities on anyone who is a benefit or pension plan fiduciary, such as corporate sponsors, directors, and officers. Severe business loss or insolvency resulting from a disaster may not affect these plans provided, of course, there is sufficient insurance coverage.

The chapter on records protection lists a number of statutes that also require organizations to have a workable disaster recovery plan.

Critical Steps in Developing a Disaster Recovery Plan

1. Identify risks, list possible threats (natural, human, accident, or equipment related) and the impact of each. Remember, the top risks of damage or destruction to computers and systems are fire and related catastrophes, mechanical breakdowns and power variations or failure, flood and water damage, and criminal acts such as vandalism, theft, and computer viruses.

2. Estimate the probability of each of the preceding for your information systems and facility.

3. Analyze your vulnerability to specific threats and the effect on information systems and business operations.

4. Prepare a list and inventory of vital resources necessary for ongoing operations. Identify critical information.

5. Examine business and legal priorities, from purchasing and contract obligations, to records management, that may be necessary for compliance or emergencies during recovery. Do a thorough review of insurance coverage—property loss, business interruption, and extra expenses.

6. Prepare a written recovery plan. Concentrate on the top hazards; in frequency and dollar loss, these are fire, power failure, and water damage. Be prepared for total disaster—assume the worst. The plan always can be scaled back for lesser emergencies. Balance cost against risk. No organization needs 100 percent redundancy for every function.

7. Make sure everyone knows what to do. From senior management responsible to setting recovery in motion to team leaders in the field, everyone must know in advance what to do during a crisis.

8. Determine the feasibility of the recovery plan. Test it periodically; evaluate and update it. If you have a hot or cold backup site, test the plan for coordination with the vendor and your suppliers.

Case Study: Handling a Freak Disaster

The 1992 flood in downtown Chicago, caused by a tunnel collapse, was dubbed by the *Chicago Tribune* as winner of "the Most Bizarre Physical Disaster Made Worse by Bureaucratic Bungling and Possible Malfeasance Contest."

A construction crew, installing pylons in the Chicago River, accidentally fractured the supporting wall of a century-old freight tunnel that runs underneath the Chicago Loop. The resulting hole released 250 million gallons of polluted water through the tunnel and into Chicago's principal business district. The freight tunnel, originally designed to deliver goods and coal and remove trash, now serves as a conduit for telephone, communications, and power cables.

The Chicago Board of Trade, four blocks from the river, had to shut down when its basement flooded. Big department stores, including Marshall Field's and Carsons, lost power when basements flooded; typically, the electrical vaults are in basements. At Carsons, the tie-lines connecting its electronic mail system went down, disrupting communication between its stores in the suburbs and out of state. Flood waters rose into basements throughout the Chicago Loop, cutting off all electrical services and forcing evacuation.

As electricity was cut off, businesses that depended on computers and telecommunications were either prepared with disaster recovery plans or suffered immediate data loss and business interruption for days or weeks.

Companies depending on computers and telecommunications lines for their primary business functions can learn a lesson from several companies who weathered the Chicago flood with minimal losses. Each had a disaster recovery plan in place and set it into operation quickly after the flood hit. These recovery plans helped the companies resume operations within hours of evacuation.

The companies had agreements for "hot sites," some located just outside of downtown Chicago, that had mainframe computers, personal computers, telecommunications lines, and office space and equipment. Comdisco Data Recovery Services (CDRS) was a provider of hot-site services and disaster recovery planning.

Electrical power surges brought down one company's mainframe. CRDS arranged for that company's circuits to be switched, if the situation warranted, to the CDRS hot site in California. As a precaution, a complete set of backup tapes was sent to the site by commercial jet.

Once this safety net was in place, the company set critical checkpoints to determine if it should fly key personnel to California and switch operations to the hot site. Fortunately, three critical points passed without further electrical problems, and the company resumed processing at its Chicago office.

At another company, the manager of Disaster Recovery Planning convened a meeting of the Disaster Recovery Steering Committee within 45 minutes after he received the first reports of water rising in the subbasements of a building adjacent to the company's main office, where the company leases space for 270 staff members. Within that same time, that building had been closed and the company had officially declared a disaster situation with CDRS. The steering committee had overall responsibility for managing the flood situation and met periodically during that week to assess and manage the changing situation.

The company's technical staff made plans to bring up their systems in a standby mode at a hot site, as a precaution against losing power in the Loop. The company maintained a staff in a state of readiness at the hot site should the power in the Loop go down.

Business losses from the flood and power outages have been estimated at over $1 billion. Flood-related claims and pending class-action lawsuits are about $400 million.

Chicago city officials claimed Great Lakes Dredge & Dock Company damaged the tunnel by negligently driving pilings into the riverbed close to the tunnel. The company claimed the city did not tell it about the tunnel and that it had failed to maintain the tunnel.

The city argued that the lawsuits belonged in state court, where it might have immunity and Great Lakes would have greater liability. Great Lakes argued that the lawsuits belonged in federal court and under the Admiralty Extension Act, where its liability is limited and it could seek indemnity from the city.

In February 1995, the Supreme Court determined that admiralty law governed the case because the alleged tort damage was committed on a navigable river by a vessel; thus Great Lakes' activity "bears a substantial relationship to traditional maritime activity" and the incident had "a potentially disruptive impact on maritime commerce."

The Supreme Court said that the circuit court was right in holding that the Admiralty Extension Act confers maritime jurisdiction over tort. This decision means that Great Lakes comes under the protection of the Limitation of Vessel Owner's Liability Act; and if the admiralty court finds that Great Lakes committed a tort, its liability could be limited to the value of the barges or tugs involved—about $630,000.

This leaves the city of Chicago with a potentially huge liability bill if it is found to have been negligent.

For more on the case, see *Jerome B. Grubart, Inc.* v. *Great Lakes Dredge & Dock Co. et al.*, and *City of Chicago* v. *Great Lakes Dredge & Dock Co. et al.*

Equipment Backup

Being prepared for equipment failure or destroyed disks will make the recovery back to normal operations much less traumatic. Large organizations are multivendor environments with many microcomputers and peripheral equipment. Standard and compatible equipment will provide users with backup processing capabilities if one machine should fail. For some organizations, it might be advantageous to warehouse sufficient units to cover critical applications needs. Large volume buyers also may find it easier to guarantee fast delivery from vendors in case of an emergency. Multivendor service providers often offer service and repair covering equipment made by many different companies.

In a one-computer organization, however, equipment failure and a long repair time could be disastrous. But standard or compatible equipment can be beneficial here, too. If the local computer store had to special order that one-of-a-kind printer, a repair can be expected to take much longer than it would if the equipment were a standard model the store's repair shop was capable of fixing. Another alternative for a small organization to pursue is an agreement with another organization in the same situation. Mutual assistance in the event of equipment malfunction can be arranged ahead of time. Local users groups can be a starting point for finding a compatible configuration.

Whatever the contingency plans for equipment may be, comparable plans for data and processing procedures also should be in place. These should include identifying what is the critical wait time for repair before manual procedures are implemented to replace those that were done using the microcomputer.

Part of good contingency planning is having adequate backups of both data and programs. Along with the backup, however, procedures should be in place for restoring data and programs so processing can continue. Before processing continues, another copy of the backup should be made so that the only copy of the data is not accidentally destroyed. Backup scheduling software programs can be used to ensure a routine is followed. Furthermore, utility programs are available that can recover seemingly lost data from diskettes. These utility programs should be purchased ahead of time and tested to make sure they can be used in an emergency.

Adequate insurance coverage for both hardware and software should not be forgotten. Also, all serial numbers and software license information should be recorded and kept in a secure location.

10

Electronic Commerce

The Electronic Funds Transfer Act (EFTA) of 1978 (Title XX, 15 U.S.C., Sect. 1693 et seq.) is the federal statute that covers a wide range of electronic funds transfers, including point-of-sale transactions and other consumer payments. If any portion of a funds transfer is covered by EFTA, the whole funds transfer is excluded from the Uniform Commercial Code Section 4A, which covers commercial, bank-to-bank transactions.

Secure On-Line Consumer Payment Systems

Commerce on the Internet has spawned a number of different ways for customers to order products and pay for them. With security perceived as a major problem in exposing credit card information, several systems have been proposed. More will come because the Comptroller of the Currency has said the private sector should lead the way in technology solutions to securing electronic transactions. One type of secure payment system takes the customer's financial information off-line, encrypts it, and never displays it live on the Internet. The most common encryption method is the Data Encryption Standard (DES); another is public key encryption; other software applications may use a Graphics User Interface (GUI) keypad, employing proprietary algorithms that randomly encrypt numbers (for example, on a credit card) and send the numbers one at a time. Another method gives "digital certificates" to on-line banks, merchants, and consumers. The certificates can include personal identification numbers (PINs), an authentication mechanism, and encryption.

Legal Support for Digital Signatures

Attorney Benjamin Wright's unpublished 1995 paper "The Legality of the PenOp Signature" supports the legality of signatures captured with pen computers, specifically Peripheral Vision's PenOp technology, a pen-based computing software component that captures and verifies signatures and links them to specific electronic documents. The PenOp is a biometric device measuring an individual's unique writing characteristics, such as pen speed, pressure, number of strokes, and other dynamic characteristics as a user writes on a computer tablet or digitizer.

The PenOp combines this analysis with encryption to create a digital record of the signer. The encrypted information also contains data on the date and time of signing.

Wright concludes that a compelling argument can be made that PenOp, when used intelligently, can allow documents to be signed in a way that is just as legally effective as the traditional method of signing. Further, adequately controlled records of the signed electronic documents are likely to be roughly as useful as evidence in court as equivalent paper documents. PenOp provides a set of security and evidence tools, including act-of-signing statistics, checksums, and a time stamp.

A broad legislative definition of a digital signature is to be found in California's Digital Signature Act. A *digital signature* is defined as an electronic identifier, created by a computer, and intended by the party using it to have the same force and effect as the use of a manual signature. Although addressing transactions in state government only, the act defines any written communication in which a signature is required or used,

> any party to the communication may affix a signature by use of a digital signature that complies with the requirements of this section [of the act]. The use of a digital signature shall have the same force and effect as the use of a manual signature if and only if it embodies all of the following attributes:
> - It is unique to the person using it.
> - It is capable of verification.
> - It is under the sole control of the person using it.
> - It is linked to data in such a manner that if the data are changed, the digital signature is invalidated.
> - It conforms to regulations adopted by the Secretary of State [of California].

Commercial Electronic Funds Transfers

The Uniform Commercial Code (UCC) sets rules and procedures that govern commercial transactions, including sales and payment methods, in most of the United States. A section of the UCC, Article 4A, establishes standards and definitions for commercial wire transfers and specifies the responsibilities of the parties involved.

UCC 4A covers funds transfers involving one or more payment orders or credit transfers; a payment order may be transmitted by telephone, fax, telex, mail, or computer. UCC 4A does not cover the following:

- Debit, check, or credit card transactions;
- Nonbank wire services;
- Any funds transfer of which any part is governed by the Electronic Funds Transfer Act of 1978 or 12 CFR Part 205, Regulation E;
- Payment orders that state a condition of payment other than value and date.

Key Terms

The following key terms occur throughout the chapter:

Credit transfer—a funds transfer where the instruction is given by the person making the payment; Article 4A governs these transfers.

Fedwire—the Federal Reserve wire transfer network; wire transfers made by Fedwire are governed by Federal Reserve Regulation J (12 CFR Part 210).

Intermediary bank—a receiving bank other than the originator's bank or the beneficiary's bank.

Payment order—an instruction by a business or the originator of the funds transfer to its bank to pay or credit an intended recipient a specific amount.

Sender—includes the customer in whose name a payment order is issued if the order is the authorized order of the customer under 4A-203 subsection (a), or it is effective as the order of the customer under 4A-203 subsection (b).

Wholesale wire transfers—transfer payments principally between businesses or financial institutions.

Where Fraud Is Most Likely to Occur

Fraud is most likely to occur in the payment order procedure. Under UCC 4A, the solutions to problems of error and fraud are spelled out. This chapter examines the key sections of the UCC 4A amendment regarding fraudulent or erroneous transfers, legal liability, and security procedures, concentrating on the section that deals with the definition of "commercially reasonable security procedures." This phrase could be called the heart of UCC 4A because it affects payment orders and establishes liability for error and fraud.

Keep in mind that UCC 4A is a model law designed for funds transfer and that it is drafted rather loosely so courts can interpret the meaning of specific clauses or sections on a case-by-case basis.

Also, UCC 4A, like the rest of the Uniform Commercial Code, is a state law; since this model law was introduced in 1991, it has been adopted by over 40 states.

The main goals of UCC 4A are to

- Reduce the risk of loss from erroneous or unauthorized payment orders;
- Create a flexible method of allocating risks and liabilities between the contracting parties;
- Establish clear and definite liability standards;
- Allow users to maintain a funds transfer system that is fast, inexpensive, and of low risk;
- Allow users leeway in devising security measures.

How Funds Transfers Operate

There is a general scenario for wire transfers: X, a debtor, wants to pay an obligation to Y. X transmits an instruction to his bank to credit a sum of money to the bank account of Y. X's bank carries out X's instruction by instructing Y's bank to credit Y's account in the amount that X requested. This series of transactions is a funds transfer. X is the originator, the instruction is a payment order, and X's bank is the originator's bank. Y is the beneficiary and Y's bank is the beneficiary's bank. In more complex transactions, there may be additional banks, called *intermediary banks*, between X's and Y's banks.

Although most, if not all, of the preceding transactions may be transmitted electronically, the actual means of transmission is not important legally. Section 4A uses the encompassing term *funds transfer*, rather than the narrower term *wire transfer*.

Electronic transfers are growing rapidly because they are more efficient, faster, and less expensive for banks than paper-based transactions. Speed, low cost, and an accompanying routine handling of electronic transactions can lead to problems, including erroneous or unauthorized payment orders. Unlike paper checks, no authorized signature is required to carry out a funds transfer.

Security Techniques and Procedures

The problems of authentication and verification of a payment order are covered under Section 4A-201, Security Procedure. Security procedure is defined as "a procedure established by agreement of a customer and a receiving bank for the purpose of (i) verifying that a payment order is that of the customer, or (ii) detecting error in transmission or the content of the payment order or communication. A security procedure may require the use of algorithms or other codes, identifying words or numbers, encryption, callback procedures, or similar security devices."

A large percentage of payment orders and communications amending or canceling payment orders are transmitted electronically and it is standard practice to use security procedures designed to assure the authenticity of the message. Security procedures also can be used to detect error in the content of messages or to detect payment orders transmitted by mistake, as in the case of multiple transmission of the same payment order. Security procedures also might apply to communications transmitted by telephone or in writing. The definition of *security procedure* limits the term to a procedure "established by agreement of a customer and a receiving bank." The term does not apply to procedures that the receiving bank may follow unilaterally in processing payment orders. The question of whether loss that may result from the transmission of a spurious or erroneous payment order will be borne by the receiving bank or the purported sender is affected by whether a security procedure was or was not in effect and whether there was or was not compliance with the procedure.

It is important that the drafters of UCC 4A used the word *procedure* and that they did not say *commercially reasonable security method* or techniques or equipment. The word *procedure* implies a process, a series of acts. Approaching *commercially reasonable security procedure* from the legal standpoint, and how it might be viewed in court, should help in determining how to set up an effective and acceptable security procedure for a funds transfer operation.

After establishing that a security procedure will be used to authenticate payment orders, the procedure must meet a standard of being "commercially reasonable." Basically, this is a question of law and worked out in an agreement between the bank and the customer. Compliance by both parties to the agreement, however, is a question of fact. Section 4A-202 and 203 provide guidance on what is meant by a *commercially reasonable security procedure.*

A security procedure is not commercially unreasonable simply because another procedure might have been better or because the judge deciding the question would have opted for a more stringent procedure. The standard is not whether the security procedure is the best available. Rather it is whether the procedure is reasonable for the particular customer and the particular bank, which is a lower standard. On the other hand, a security procedure that fails to meet prevailing standards of good banking practice applicable to the particular bank should not be held to be commercially reasonable.

UCC 4A is not vague, it is only loose in its description of what is required in the way of security. But the point here is not to dwell on a failure in the law to define in concrete terms a security technique that is commercially reasonable. Instead, the statute says that interpretation of the standard is a "question of law," not one of engineering.

The type of payment order and how it is transmitted also must be considered. In a wholesale wire transfer, "each payment order is normally transmitted electronically and individually. A testing procedure will be individually applied to each payment order. In funds transfers to be made by means of an automated clearing house many payment orders are incorporated into an electronic device that is physically delivered. Testing of the individual payment orders is not feasible. Thus, a different kind of security procedure must be adopted to take into account the different mode of transmission."

The Courts and UCC 4A

UCC 4A is relatively new, and case law is scarce. However, we can try to think of how a court could look at and interpret various sections of UCC 4A. This is not a mere exercise in conjecture; it is an attempt to forge a proactive stance in the face of possible litigation.

In adjudicating a case involving UCC 4A, the court surely will look at two things: the process of decision making of the parties and the presented evidence. Let us take these one at a time.

Again, fraud is most likely to occur in the initiation of a payment order and that it where identification and verification of the sender is critical. What procedures were established, how were they established, what was the process, how was an agreement reached, and is the agreement in writing? These are the important questions to cover. Section 4A-203 states:

> Some person will always be identified as the sender of a payment order. Acceptance of the order by the receiving bank is based on a belief by the bank that the order was authorized by the person identified as the sender . . .
>
> . . . Given the large amount of the typical payment order, a prudent receiving bank will be unwilling to accept a payment order unless it has assurance that the order is what it purports to be. This assurance is normally provided by security procedures . . .
>
> . . . The receiving bank may be required to act on the basis of a message that appears on a computer screen. Common law concepts of authority of agent to bind principal are not helpful. There is no way of determining the identity or the authority of the person who caused the message to be sent. The receiving bank is not relying on the authority of any particular person to act for the purported sender . . . Rather, the receiving bank relies on a security procedure pursuant to which the authenticity of the message can be "tested" by various devices which are designed to provide certainty that the message is that of the sender identified in the payment order. In the wire transfer business the concept of "authorized" is different from that found in agency law.
>
> [Subsection (b) of Section 4A-202] is based on the assumption that losses due to fraudulent payment orders can best be avoided by the use of commercially reasonable security procedures . . . A receiving bank needs to be able to rely on objective criteria to determine whether it can safely act on a payment order. Employees of the bank can be trained to "test" a payment order according to various steps specified in the security procedure. The bank is responsible for the acts of these employees . . . If the fraud was not detected because the bank's employees did not perform the acts required by the security procedure, the bank has not complied. Subsection (b)(ii) also requires the bank to prove that it complied with any agreement or instruction that restricts acceptance of payment orders issued in the name of the customer.

Once the order has been accepted, the bank is obligated to carry out the order: "acceptance means that the receiving bank has executed the sender's order and is obliged to pay the bank that accepted the order issued in execution of the sender's order" (Section 203 (b)).

Liability

The question of who is to blame or who is liable for fraudulent or errone-ous payment orders hinges mainly on security procedures. For example, a customer may choose to use a cheaper, but high-risk security proce-dure. He or she may do this and seemingly assume the risk of loss. The bank, however, is free of risk only if it first offered the customer a com-mercially reasonable security procedure as an alternative and the agree-ment to go with the cheaper procedure has been expressed in writing.

"If a commercially reasonable security procedure is not made available to the customer, subsection (b) does not apply. The result is that subsection (a) applies and the bank acts at its peril in accepting a payment order that may not be authorized."

Again, all this relies on a written agreement: ". . . the receiving bank is not required to follow an instruction that violates a written agreement."

Interloper Fraud and Liability

A breach of commercially reasonable security procedures, and atten-dant blame, "requires that the person committing the fraud have knowledge of how the procedure works and knowledge of codes, iden-tifying devices, and the like. That person may also need access to trans-mitting facilities through an access device or other software. . . . This confidential information must be obtained either from a source control-led by the customer or from a source controlled by the receiving bank."

The customer's responsibility is "to supervise its employees to as-sure compliance with the security procedure and to safeguard confi-dential security information and access to transmitting facilities so that the security procedure cannot be breached."

> If the customer can prove that the person committing the fraud did not obtain the confidential information from an agent or for-mer agent of the customer or from a source controlled by the cus-tomer, the loss is shifted to the bank.

The customer, then, is not liable if it can prove that the bank or an outsider, for example, a hacker, compromised its security.

Liability and Proof

Liability for accepting a fraudulent payment can occur for the bank if its employees fail to "test" a payment order according to various steps specified in a security procedure. The bank is required to "prove that it accepted the payment order in good faith and in compliance with the security procedure. If the fraud was not detected because the bank's employees did not perform the acts required by the security procedure, the bank has not complied." Further, the bank must prove that it complied "with any agreement or instruction that restricts acceptance of payment orders issued in the name of the customer."

Evidence Gathering

Evidence of fault or fraud and the cause of the loss usually will be developed through a criminal investigation as well as an internal investigation of the bank. In addition, there may be an investigation by bank examiners. The bank's customer "will have access to evidence developed in these investigations and that evidence can be used by the customer in meeting its burden of proof."

Unauthorized Payment Orders

Unauthorized payment orders normally are the responsibility of the customer, in that "the customer has a duty to exercise ordinary care to determine that the order was unauthorized after it has received notification from the bank, and to advise the bank of the relevant facts within a reasonable time not exceeding 90 days after receipt of notification."

Summary

A security procedure is an agreed-upon procedure between a bank and its customer for verifying that a payment order is that of the customer and that any error in the electronic transmission or the content of the payment order or communication is detectable. Funds transfer security techniques may involve the use of encryption, codes, callback systems, or similar identification and verification devices.

Funds transfer security procedures may include

- Limiting the amount of any transfer;
- Limiting the number of authorized beneficiaries;
- Making transfers payable only from an authorized account;
- Prohibiting any transfer that exceeds specific credit limits or account balances;

- Initiating and verifying wire transfer transactions by different operators;
- Using built-in system access controls that compare identifications of initiator and verifier;
- Repeating the same procedures for incoming funds transfers to detect and prevent diversion to an unauthorized account;
- Testing security procedures.

Devices or software may also have the capability to

- Verify that payment and account numbers match up;
- Avoid duplicate payment orders by identifying each order sequentially and with a unique code;
- Detect transmission errors.

Commercially reasonable security depends on

- The wishes of the customer;
- Security procedures used by similar banks;
- Alternative security procedures offered to the customer;
- Circumstances of the customer known to the bank;
- Frequency of payment orders issued by the customer.

To ensure the accuracy, auditability, and recoverability of electronic records:

- The electronic document capture must be well-defined, controlled, and auditable to support a claim to authenticity;
- Final-form electronic records must be stored in a secured repository;
- A records contingency recovery program must be created and tested.

Audit considerations include

- Having a written agreement between contracting parties;
- Offering the customer commercially reasonable security procedures;
- Documenting compliance with security procedures;
- Documenting message authentication of transactions;
- Having document acceptance and notification policies and procedures;

- Having a disaster recovery plan for computer and network systems and records storage and recovery.

Conclusion

How a commercially reasonable security procedure is chosen and implemented is a decision-making process by the bank and the customer. The procedure, including various security methods and equipment, will be embodied in a written agreement between the bank and the customer.

The court will look at both the written agreement and the decision-making process that led to the agreement. Both the process and the agreement should reflect the security measures, and alternatives, offered the customer, the desires of the customer, and his or her obligations.

Finally, it is critical to have detailed record keeping of written customer agreements, agreement modifications, acceptance policies, and transactions as well as an audit trail of security procedures.

The American National Standards Institute (ANSI) has security guidelines for wire transfers and how to audit security procedures.

It also may be a wise proactive policy to have on file a list of technical experts on your security system who can testify in court cases on message authenticity and data integrity.

Electronic Data Interchange

Electronic data interchange (EDI) is the automated exchange of structured business data, such as invoices, purchase orders, and other documents and forms via computer between businesses. If payment information is exchanged—credit or debit instructions—through the banking systems' automated clearing houses or certain value added networks (VAN), then it is financial EDI. Usually these are separate exchanges of financial and nonfinancial data, depending on networks, VANs, and banks.

This type of information exchange is based primarily on a contract: a set of promises that, if broken, provide a legal remedy. This means that trading partner agreements can include descriptions of forms and documents, payment terms, and delivery schedules, as well as specific control, security, and audit measures. The American Bar Association has a Model Trading Partner Agreement, which defines the terms of an acceptance and allows for adoption by parties of electronic signatures ("symbols or codes which are to be affixed to or contained in each document transmitted by such party"), encryption, and other security measures.

All contracts for the sale of goods over $500 come under U.C.C., Section 2-201, the Statute of Frauds. This section requires the contract to be in a "writing" and that it be "signed" by the party. To meet the requirements of 2-201, the writing must evidence a contract for the sale of goods, be signed in a way that authenticates and identifies the party to be charged, and specify the quantity. A complete signature to authenticate a writing is not necessary. "Authentication may be printed, stamped, or written; it may be by initials or by thumbprint. It may be on any part of the document and in appropriate cases may be found in a billhead or letterhead."

As in funds transfers, security measures can support the legal enforcement of contracts and agreements by ensuring the authentication and integrity of the communication. However, courts will look at the claimed and to-be-proven effectiveness of security techniques and equipment, as is the case with security under UCC 4A.

Record Keeping for EDI

EDI record keeping, backup, storage, and recoverability should follow those described for other electronic records. The purpose is to ensure that EDI records will provide legal proof of transactions should a legal dispute arise. Another reason is to satisfy government records requirements. In EDI, when to capture and store data is critical but often system or transaction specific. One also should consider keeping a complete trade data log of all transfers sent and received. The goal is getting the most accurate reproduction of an original record that can be made; in short, for business and legal purposes, a reliable and provable final-form record that will be in a secured electronic repository.

11

E-Mail Policy Guide

E-Mail Maledicta: Messages That Wound, Embarrass, and Lead to Lawsuits

E-mail is seen as analogous to a private postal system, a telephone, a bulletin board for tacking up messages, or the World Wide Web with near instantaneous broadcasting capabilities. An e-mail system can have all these characteristics plus user convenience, informality, and spontaneity.

At times, however, some e-mail users send messages that are demeaning, abusive, discriminatory, or simply "untouched by human thought." Lawyers love to find these uncensored messages on e-mail systems—they refer to them as *hot documents* because such potentially incriminating statements can provide an opening to possible liability and litigation.

Operationally, e-mail systems provide not only message transmission but may have a time/date stamp and a record of who received the message and when—a logging and tracking of messages through a company and between companies or individuals. There also may be a record of the message distribution list as well as automatic archiving of messages. In short, a complete audit trail of who is sending what to whom—and a road map to finding incriminating statements, or tracing a decision-making process to determine accountability.

An employer is vicariously liable for the tortious acts of an employee who is acting within the scope of his or her employment—this may include "acts" committed via e-mail. An emerging legal concept is the tort of negligent supervision, or failing to supervise an offending

125

employee. Organizations are faced with both lessened standards of liability and mandates to report possible regulatory, civil, or criminal misconduct of their employees or even by the organizations themselves.

This chapter is designed to be a guide for employers, managers, supervisors, and employees on how to avoid language and acts via e-mail systems that could infer defamation, discrimination, harassment, invasion of privacy, negligence, or other liability against individuals, the organization, or its agents.

Although legal cases specific to e-mail are scarce (so far), we will examine those that are relevant along with cases that illuminate statutes on sexual harassment, privacy, and the use of harmful language. Drafting an e-mail policy plus sample policies on system use and evidence are included in this chapter.

Throughout this chapter the focus will be on e-mail; however, many of the problems discussed and proposed solutions also could apply to bulletin boards and Web sites. These systems also transmit information to the public, thus affecting the potential liability exposure of the organization.

Flaming On-Line Can Get You a Libel Suit

A message, or "flame" in on-line parlance, was posted on Prodigy's Money Talk bulletin board by a subscriber who accused the investment banking firm of Stratton Oakmont of "major criminal fraud" and "100 percent fraud" in an initial public stock offering. Stratton's president was described as a "soon to be proven criminal."

A New York state court ruled in *Stratton Oakmont Inc.* v. *Prodigy Services Co.* that Prodigy could be held liable for libelous statements posted by a subscriber. Stratton Oakmont sought $200 million in damages.

The trial judge ruled that Prodigy is a publisher with the responsibilities that go with the content it publishes; the on-line service should not be considered simply a "common carrier" with lesser responsibilities for content. The judge also pointed out that Prodigy had issued content guidelines, that Money Talk had an editor who was to delete messages that violated guidelines, used a software screening program to weed out offensive language, and had a bulletin board standards group.

The judge said that it was Prodigy's "own policies, technology and staffing decisions which have altered the scenario and mandated a finding that it is a publisher. . . . Prodigy held itself out to the public and its members as controlling the content of its computer bulletin boards."

And, "to gain the benefits of editorial control, [Prodigy] has opened itself up to a greater liability than CompuServe and other computer networks. . . . "

This case raises many issues such as the definitions of *slander* and *libel* in on-line systems, whether electronic publishers are defined by the degree of editorial control they exercise, and whether controls of various kinds create an enforceable liability.

Defamation Defined

Wrongful or negligent disclosure of private or embarrassing facts usually requires such information to be communicated to more than one person. Any disclosure of false information could lead to a defamation suit. Defamation has two types of communication: defamation via print, writing, pictures, or signs is called *libel*; *slander* is defamation by speech. Both are the communication of false information to a third party that injures a person's or a business's reputation—causing bad opinion, public hatred, ridicule, or disgrace.

Other elements of defamation include the reasonable identification of the defamed person; damage to the person's reputation; if the defamation refers to a public figure or is a matter of public concern, it must be proven that the defamatory language was false and that it was communicated knowingly or with a reckless disregard as to the truth or falsity of the information.

The basic defenses to defamation are that the facts of the statement are provably true and that a privilege can be invoked. Privilege can be absolute, which is reserved for government officials, such as judges and legislators, and the content of most public records. The press has a qualified or limited privilege to report on matters of public interest that might go unreported. This qualified privilege can be lost if the information is in error and malice can be shown.

A Hypothetical Case of What Not to Do

An employer suspects an employee is stealing and fires him. As a warning to other employees, the employer sticks a notice on the company bulletin board stating that the employee has been fired for stealing company property.

Could the employer be liable in defamation per se? You bet.

What if the notice had been posted on the company's e-mail system? It is the same as the bulletin board.

This based on an actual case. The employee was never tried or convicted of theft, and there was no proof a theft ever took place. His ex-employer was found liable of damaging the employee's reputation, proof of which was the former employee's many rejections for employment.

E-Mail Used as Evidence in Discrimination Suit

Karen Strauss, then an employee of Microsoft Corp., received an e-mail message sent by her supervisor containing a satirical essay entitled "Alice in UNIX Land." The supervisor also called another woman in the office the "Spandex Queen" and sent a male staffer an e-mail message that contained a parody of a play called "A Girl's Guide to Condoms," which the staffer later forwarded to Strauss.

Evidence of the supervisor's office behavior, including e-mail messages with sexual undertones, was admissible in a gender discrimination suit brought by Strauss, now a former employee of Microsoft. The plaintiff argued that gender played a part in a Microsoft supervisor's failure to promote her. Even though the e-mail messages did not prove that gender played a part in the failure to promote, they were relevant to the discrimination issue.

Microsoft wanted the e-mail messages and other evidence ruled inadmissible because it was irrelevant, prejudicial to Microsoft, and would confuse and mislead the jury. Microsoft characterized the supervisor's remarks and e-mail messages as attempts at humor and not directly connected to its promotion and termination decisions.

The Court found this argument meritless. The Court said Rule 401 of the Federal Rules of Evidence provides that evidence is relevant if it makes "the existence of any fact that is of consequence to the determination of the action more probable or less probable than it would be without the evidence." Evidence of the supervisor's inappropriate office comments and e-mail messages, when viewed in light of the plaintiff's other evidence of pretext, could lead a reasonable jury to conclude that Microsoft's proffered reason for failing to promote Strauss is not the true reason for its actions. (See *Strauss* v. *Microsoft Corp.*, U.S. District Court, Southern District of New York.)

Creating an Abusive or Hostile Work Environment

The following Supreme Court case, *Harris* v. *Forklift Systems,* does not involve e-mail; however, it is a recent and landmark case on sexual harassment. Its scope and the definitions and limits of abuse, including abusive language, described in the case are relevant for every workplace.

Teresa Harris worked as a manager at Forklift Systems, Inc., an equipment rental company. Throughout her two-year employment at Forklift, Charles Hardy, the firm's president, often insulted Harris and made her the target of unwanted sexual innuendoes. Hardy, in front of other employees, called Harris "a dumb ass woman." Another time, he suggested the two of them "go to the Holiday Inn to negotiate [Harris's] raise"; and when Harris was arranging a deal with one of Forklift's customers, Hardy asked her, "What did you do, promise the guy . . . some [sex] Saturday night?"

Harris quit her job, then sued Forklift, claiming that Hardy's conduct had created an abusive work environment for her because of her gender. A lower court found that, although Hardy's acts and comments were offensive, they were not "so severe as to be expected to seriously affect [Harris's] psychological well-being." The U.S. Supreme Court took the case to "resolve a conflict among the circuits on whether conduct, to be actionable as 'abusive work environment' [under Title VII of the Civil Rights Act of 1964] must 'seriously affect [an employee's] psychological well-being' or lead the plaintiff to 'suffer injury.'"

Title VII makes it "an unlawful employment practice for an employer . . . to discriminate against any individual with respect to his compensation, terms, conditions, or privileges of employment, because of such individual's race, color, religion, sex, or national origin." The Supreme Court has said that this language is meant to cover the entire spectrum of disparate treatment in employment, which includes requiring people to work in a discriminatorily hostile or abusive environment. However, "mere utterance of an . . . epithet which engenders offensive feeling in an employee" does not sufficiently affect the conditions of employment to implicate Title VII.

Title VII does come "into play before the harassing conduct leads to a nervous breakdown. A discriminatorily abusive work environment, even one that does not seriously affect employees' well-being, can and often will detract from employees' job performance, discourage employees from remaining on the job, or keep them from advancing in their careers." Although Title VII bars conduct that would seriously affect a person's psychological well-being, the statute is not limited to such conduct. In the Harris case, the Court concluded that whether an environment is hostile or abusive can be determined only by looking at all the circumstances. And, while the employee's well-being is relevant, psychological harm, although a relevant factor and one that can be taken into account, no single factor is required. "So long as the environment would reasonably be perceived, and is perceived, as hostile or abusive, there is no need for it also to be psychologically injurious."

The Supreme Court found in Harris's favor, ruling that workers suing their employers for sexual harassment did not have to show that they suffered psychological injury.

EEOC Guidelines on Workplace Harassment

The EEOC has issued guidelines on harassment, which may include the following conduct relating to race, religion, gender, national origin, age, or disability: epithets, slurs, negative stereotyping, threats, hostile acts, and denigrating or hostile written or graphic material posted or circulated in the workplace.

An employer, notes the EEOC, is liable for the conduct of its employees and that of its agents and supervisory employees whenever the employer knows of or should know of harassment by its employees or agents and fails to take immediate and appropriate corrective action. Further, also liable are employers that fail to implement an explicit policy against harassment that is clearly and regularly communicated to employees and employers that fail to establish an accessible procedure allowing employees to make harassment complaints known to appropriate officials.

Wicked Words

Generally, bad or abusive words are of three types: profane, obscene, or insulting. In law, these words can create a conflict between an individual's First Amendment right of free speech and restrictions imposed by statutes or employers. Cases nearly always are situation specific; however, the courts have offered some guidance. The Supreme Court in *Chaplinsky* v. *New Hampshire* stated:

> There are certain well-defined and narrowly limited classes of speech, the prevention and punishment of which have never been thought to raise any constitutional problem. These include the lewd and obscene, the profane, the libelous, and the insulting or "fighting" words—those which by their very utterance inflict injury or tend to incite an immediate breach of the peace . . . Such utterances are no essential part of any exposition of ideas, and are

of such slight social value as a step to truth that any benefit that may be derived from them is clearly outweighed by the social interest in order and morality.

And in *R.A.V.* v. *City of Saint Paul*, Justice Scalia used sexual harassment as a proscribable class of speech: "Thus, for example, sexually derogatory 'fighting words' among other words, may produce a violation of Title VII's general prohibition against sexual discrimination in employment practices. Where the government does not target conduct on the basis of its expressive content, acts are not shielded from regulation merely because they express a discriminatory idea or philosophy."

The Tort of Outrage

An employee who is the target of abusive verbal or e-mail messages, such as racial slurs or epithets, could bring a tort of outrage claim against the employer. According to this tort law, "One who by extreme and outrageous conduct intentionally or recklessly causes severe emotional distress to another is subject to liability for such emotional distress . . . "

"Intentional infliction of emotional distress" also could come from threats of violence. Note that is not necessary to show that someone who threatens another with physical harm is likely to commit acts of violence. There is simply no certain correlation between threats and violent acts. While a person might fit a "violence-prone profile," profiles generalize personality traits denominated from a large group; their use in assessing individual risks is limited. However, knowing that an employee uses threatening and violence-laden language should alert an employer to a general duty to foresee the possibility of violence.

Disability Defined by the EEOC

The Equal Employment Opportunity Commission (EEOC) has released definitions of what constitutes a disability under the Americans with Disabilities Act of 1990 (ADA). The definition is tailored to the purpose of eliminating discrimination prohibited by the ADA. A determination of whether a charging party has a "disability" turns on whether he or she meets the definition of that term.

A charging party has a disability for purposes of the ADA if he or she

1. Has a physical or mental impairment that substantially limits a major life activity
2. Has a record of such an impairment, or
3. Is regarded as having such an impairment.

A charging party must satisfy at least one of these three parts of the definition to be considered an individual with a disability. When determining whether a charging party satisfies the definition of disability, remember that the concepts of "impairment," "major life activity," and "substantially limits" are relevant to all three parts of the definition of disability. Also remember that the disability determination should be made without regard to the availability of mitigating measures. Further, certain conditions are specifically excluded from the definition of disability.

Some Disability-Related Terms Defined

An *impairment* is a physiological disorder affecting one or more of a number of body systems or a mental or psychological disorder. The following conditions are not impairments: environmental, cultural, and economic disadvantages; homosexuality and bisexuality; pregnancy; physical characteristics; common personality traits; normal deviations in height, weight, or strength.

Major life activities include caring for oneself, performing manual tasks, walking, seeing, hearing, speaking, breathing, learning, and working. Other examples include sitting, standing, lifting, mental and emotional processes such as thinking, concentrating, and interacting with others.

Substantially limits refers to an impairment that prohibits or significantly restricts an individual's ability to perform a major life activity as compared to the ability of the average person in the general population to perform the same activity.

The determination of whether an impairment substantially limits a major life activity depends on the nature and severity of the impairment, the duration or expected duration of the impairment, and the permanent or long-term impact of the impairment.

An impairment substantially limits an individual's ability to work if it prevents or significantly restricts the individual from performing a class of jobs or a broad range of jobs in various classes.

EEOC Guidelines on Preemployment Inquiries Under ADA

The EEOC has issued guidelines covering the scope of permissible interview questions under the ADA. Specific questions are regarded as "disability related" and, therefore, are prohibited at the initial job interview and preoffer stages.

Examples of Disability-Related Questions

The following are examples of questions that could reveal disability-related information:

1. Do you have AIDS? Do you have asthma?
2. Do you have a disability that would interfere with your ability to perform the job?
3. How many days were you sick last year?
4. Have you ever filed for workers' compensation?
5. Have you ever been injured on the job?
6. How much alcohol do you drink each week? Have you ever been treated for alcohol problems?
7. Have you ever been treated for mental health problems?
8. What prescription drugs are you currently taking?

Using Bias-Free Language

Learn to expurgate demeaning, dehumanizing, and patronizing words from your communications. Use words that are free from bias and stereotypes.

Aim for clarity and accuracy. It is better to refer to people with disabilities, rather than the disabled; to a person with a specific disability, for example, as blind, deaf, or has epilepsy, or walks with a cane or uses crutches.

Where possible, avoid using the male pronoun when you want to refer to both men and women—either eliminate it, replace it with a neutral article (*the* or *its*), recast the sentence in a passive voice, or rewrite the sentence to eliminate unnecessary pronouns. Rewriting often does the best job since it forces a discipline toward precision.

E-Mail Policy Considerations

To control and limit the misuse of its e-mail system, an organization must develop a policy and set of procedures that cover the following:

- Network and e-mail systems as well as connections to on-line services;
- The ownership and purposes for which e-mail is to be used;
- That there is to be fair, nondiscriminatory treatment of all employees;
- That e-mail messages that are obscene, offensive, or project discrimination, harassment, or any form of abuse will not be tolerated;
- Enforcement and penalties for misuse;
- A preventive law awareness program for e-mail users.

The organization must deal with the presence of a discrete expectation of privacy with respect to e-mail. The organization should promulgate a clear statement on e-mail privacy designed to eliminate an employee expectation of privacy and assert employer ownership, authority and oversight of the e-mail system. Employee usage of the e-mail system is at the employer's discretion. The employee uses the e-mail system with rules and restrictions that are designed to protect the employer's property and serve legitimate business interests, such as the employer's need to identify messages that might compromise the employer's legal interests.

Policies, rules, and actions should clarify that individual access to the e-mail system is via key or password controlled and administered by security or network administrators; that the network may be monitored and message contents filtered (a software filter program can block some obscene, racist, and sexual material on e-mail); and, that e-mail files may be searched and seized at any time and for any reason.

The Electronic Communications Privacy Act's (ECPA) prohibitions on e-mail system monitoring and disclosure exempt business uses, specifically allowing: "An officer, employee, or agent of a provider of wire or electronic communication service, whose facilities are used in the transmission of a wire communication, to intercept, disclose, or use that communication in the normal course of his employment while engaged in *any activity which is a necessary incident to the rendition of his service or to the protection of the rights or property of the provider of that service*" (emphasis added).

Another clause of the ECPA allows interception of communications if the "electronic device" used for interception is "any telephone or telegraph instrument, equipment or facility, or component thereof, furnished to the subscriber or user *in the ordinary course of business* and being used by the subscriber or user in the ordinary course of its business or furnished by such subscriber or user for connection to the facilities of such services and used in the ordinary course of its business" (emphasis added). Presumably, monitoring and filtering software would meet the definition of *component thereof.*

The ECPA covers only the interception of electronic communications transmitted via common carrier.

The outline of any policy may be reviewed or entirely written by legal counsel. Certainly, the final draft should be reviewed and approved by the responsible company officer and counsel.

Sample E-Mail Policy Outline and Content

1. To cover external and internal corporate e-mail system:
   a. Microcomputers, terminals, and networks;
   b. Messages, drafts, records, documents, and other information on the e-mail system, including backup media and storage.
2. Statement of content:
   a. The sole purpose of the corporation's e-mail system is to assist in conducting the business of the enterprise;
   b. All computers and communications equipment and facilities, including e-mail, and the data and information stored on them, are and remain at all times business property of the corporation and are to be used for business purposes only.
   c. It is the goal of the corporation to maintain a work environment for all its employees that is absent disparate treatment, and it will not tolerate abuse or discrimination against candidates or existing employees.
   d. The corporation's e-mail system may not contain messages having language or images that may be reasonably considered offensive, demeaning, or disruptive to any employee or create a discriminatorily hostile or abusive work environment. Such e-mail message content would include, but would not be limited to, sexually explicit comments or images, gender-specific comments, racial epithets and slurs, or any comments or images that would offend someone based on their race, color, sex, religion, national origin, age, physical or mental disability, status as a veteran, or sexual orientation.
   e. The corporation reserves the right to monitor all e-mail message content.

   f. Any views expressed by individual employees in e-mail messages are not necessarily those of the corporation.

   g. The corporation shall establish a systematic e-mail message, records, and document retention and destruction program designed to be effective in meeting the legitimate business needs and legal obligations of the corporation.

   h. Violation of corporate policies by employees will invoke disciplinary measures up to and including termination of employment.

   i. This policy will be reviewed periodically and updated in light of new legal developments and corporate experiences.

Promulgation of the E-Mail Policy

1. Disseminate copies of the policy to all those whose conduct it is to govern.
2. Be sure it is written clearly and is easily understood.
3. Have each recipient sign off on it, with initials and the date.
4. It can be communicated via
    a. The employee handbook (maintain a log of employees who receive the handbook);
    b. Posting on employee bulletin boards;
    c. Physical distribution of printed text;
    d. A message in the e-mail system (consider a "splash screen" with a policy message or warning on users' computer screens that will flash a reminder whenever an employee logs onto the e-mail system).

Violations of Policy

Enforcement and sanctions against policy violations should

1. Be consistent in application;
2. Provide disciplinary mechanisms for
    a. Illegal conduct,
    b. Unethical conduct,
    c. Failure to detect an offense,
    d. Failure to report an offense;
3. Define conduct that is grounds for termination of employment;
4. Ensure disciplinary measures do not conflict with employment laws;
5. Ensure termination action does not conflict with the personnel policy manual (consult with legal counsel on the termination action).

E-Mail Evidence

In March 1993, Siemens Solar Industries sued Atlantic Richfield Company (Arco) for fraud in the sale of its solar energy unit. The suit revealed evidence of e-mail messages between Arco employees that indicated Arco sold the unit knowing it would not be commercially viable.

In the dispute between Symantec Corporation and Borland International, Inc., over allegedly stolen trade secret marketing plans, e-mail messages were revealing evidence.

And, e-mail information (thought to have been erased) proved critical in the Iran-Contra scandal of the Reagan and Bush administrations.

Hot Documents and Road Maps

In these and other cases, electronic mail has become a valuable source of evidence in lawsuits. In pretrial maneuvers, e-mail now often is seen as a prime source of uncensored statements, producing the electronic "hot document" that can provide the opening trail to litigation and potential liability.

In the Arco case, e-mail evidence was a series of exchanges between employees. The Borland suit looked at the record of messages kept by the e-mail network. In the Iran-Contra case, backup tapes provided the evidence.

Again, keep in mind, e-mail systems offer logging and tracking of messages flowing through a company and between companies. A complete audit trail of who is sending what to whom makes it easier to find incriminating statements or trace a decision-making process to determine accountability.

Evidence Recovery

Another critical point about e-mail is that data and messages supposedly erased from a hard drive or magnetic backup media may be restored by various data recovery or forensic methods. Magnetic media is normally rewritable; deletion, in DOS, erases the pointers on the file and leaves the information in the disk's clusters.

To erase data, a file cleaning program or disk reformatting usually is necessary.

Message Encryption and Privacy

Selection of an encryption device will provide a reasonable measure only of message privacy. Encryption does not mean that the message has been erased after sending; encrypted data is no different than other data written to disk.

E-Mail Policy Considerations

To control and limit the disclosure of e-mail information, corporations and other organizations must develop a policy and set of procedures that cover the following:

- The purposes for which e-mail is to be used;
- The enforcement and penalties for misuse;
- The purging of e-mail files;
- An e-mail records retention program;
- The legal consequences of destruction or loss of e-mail evidence;
- A preventive law awareness program for e-mail users.

Policy Outline

The following outline highlights the more significant elements to examine in considering an e-mail information control policy. It is intended to provide a framework for marshaling your thoughts and ideas for creating a company-specific policy and set of procedures.

Although the policy outline has elements common to compliance programs, its major thrust is implementing a proactive strategy of controlling access to corporate information that could be used in litigation. This preventive law approach stresses the importance of a legally sound document management program and of making employees aware of how their use of e-mail could have legal consequences.

Policy Statement and Oversight Responsibility

This is issued by the corporation's chief executive officer.

1. Statement content outline:
   a. The corporation's e-mail system is considered business property and is to be used for business purposes only;
   b. The corporation reserves the right to monitor all e-mail message content;

    c. Any views expressed by individual employees in e-mail messages are not necessarily those of the corporation;

    d. The corporation shall establish a systematic e-mail message, records, and document retention and destruction program designed to be effective in meeting the legitimate business needs and legal obligations of the corporation;

    e. This policy will be reviewed periodically and updated in light of new legal developments and corporate experiences.

2. Oversight responsibility for implementing this policy is to include high-level personnel; for example,

    a. Chief financial officer,

    b. Legal counsel,

    c. MIS director,

    d. Chief information officer,

    e. Records manager,

    f. Director of the internal audit.

Communication of Policy

1. Disseminate copies of the policy to all those whose conduct it is to govern.
2. Be sure it is written clearly and is easily understood.
3. Have each recipient sign off on it, with initials and the date.

Policy Implementation and Procedures

The policy should cover the entire external and internal corporate e-mail system:

1. Microcomputers, terminals, and networks;
2. Messages, drafts, records, documents, and other information on the e-mail system, including backup media and storage.

Determine the relevance of all material on the e-mail system:

1. Is it vital to corporate operations?

2. Has it potential legal or regulatory importance?

3. Is it required by law or regulation to be part of a document retention program?

4. Is it material that can be purged in the course of business operations (cost controls, expense reductions, lack of storage space, etc.)?

The materials to be retained or purged, and on what schedule, should be determined by advice from

1. Legal counsel;
2. Records management director;
3. Division manager, department head, or group supervisor;
4. Chief financial officer or controller;
5. Chief information officer or MIS director.

Material to be retained should be

1. Sorted, indexed, and stored according to records retention requirements;
2. Vital operational material to be handled by department manager, financial officer, or chief information officer.

Material to be purged should be permanently erased or otherwise destroyed, including any copies or backups.

Purging of e-mail materials shall cease upon receipt of notice of imminent legal action, receipt of subpoena, discovery motion, indictment, or other legal notice, under these conditions:

1. Suspension notification will be from legal counsel.

2. Any corporate document or computer file purging program, including e-mail, that possibly could involve discoverable material will be suspended.

3. Depending on the discovery request, an organized search for e-mail material will be conducted.

4. All e-mail material collected will be screened for response to discovery and for privilege protection.

5. E-mail material will be identified and marked, indexed, and duplicated.

The retention and destruction program may resume

1. On notification of legal counsel;
2. On compliance and lifting of discovery motion or cessation of litigation.

Violations of Policy

Enforcement and sanctions against policy violations should

1. Be consistent in application;

2. Provide disciplinary mechanisms for illegal conduct, unethical conduct, and failure to detect and report an offense;

3. Define conduct that is grounds for termination of employment;

4. Ensure that disciplinary measures do not conflict with employment laws;

5. Ensure that termination action does not conflict with the personnel policy manual (consult with legal counsel on termination action).

Audits of Program

The purpose of an audit program is to monitor compliance with policy directives and procedures. To best accomplish this,

1. Set the frequency and timing of audits.
2. Conduct audits by an internal audit department.
3. Focus on the formal and informal management controls, and assess their effectiveness.
4. Thoroughly review all controls in each area.
5. Document and report the audit's findings.

Education and Awareness Program

State the reasons for a corporate e-mail policy.

1. Explain how e-mail should be used by the corporation's personnel (personal messages vs. business information).

2. Create an awareness of how certain uses of e-mail can affect future investigations, legal proceedings, and litigation.

3. Explain that the need for a strategy of controlling access to corporate information is necessary in today litigious environment.

4. An effective program of e-mail records retention and destruction is part of this strategy.

Explain the connection between e-mail messages and discoverable evidence:

1. Explain how certain words, phrases, and messages, often seemingly innocuous on e-mail, could lead to accusations of slander, defamation, or discrimination.

2. Explain how e-mail messages have been used as discoverable evidence in recent and varied cases:
   a. How such evidence, though seemingly minor, provided an opening, a trail leading to further incriminating evidence and litigation;
   b. Deleted e-mail data may be restored by various data recovery or forensic methods;
   c. Encrypting messages assures reasonable privacy, but encrypted information may still be recovered.

3. Identify the types of offenses the corporation must prevent; examine
   a. Records of recent complaints, labor or employee disputes, and corporate litigation;
   b. Issues identified by key personnel;
   c. The nature of the products or services offered;
   d. Experience of other businesses of similar size and e-mail systems.

4. Provide behavioral descriptions of wrongdoing and unethical behavior.

5. Explain in a clear practical manner and illustrate by example
   a. Illegal actions;
   b. What the law proscribes;
   c. The consequences for violation of the law.

Explain how the corporation will gain access to information and evidence:

1. The discovery plan will focus on
   a. Determining sources of information, location, and possible content;
   b. Types of records or documents, how maintained and circulated;
   c. Information that will support or establish the allegations;
   d. How long it takes to get the information.
2. The techniques of discovery include
   a. Document production requests;
   b. Interrogatories;
   c. Oral depositions;
   d. Subpoenas, types and purposes;
   e. Search warrants;
   f. Civil investigative demands.

3. Different disclosure and reporting requirements are included under
   a. Voluntary disclosure agreements or programs;
   b. The duty to report laws and regulations;
   c. Compliance programs;
   d. Parallel proceedings.
4. E-mail information is significant in litigation:
   a. The importance of e-mail in the discovery process;
   b. The disclosure of inculpatory information;
   c. Its relevance to potential litigation.
5. The destruction of evidence may have varying consequences:
   a. Evidence destroyed once litigation is pending or imminent;
   b. Failure to preserve evidence reasonably known to be relevant or discoverable;
   c. Evidence destroyed deliberately or recklessly, inadvertently and without malice;
   d. Selective destruction of files and information prior to litigation.
6. Explain the meaning of a notice to cease document destruction: notification from legal counsel of receipt of discovery order, complaint filed, indictment issued, or start of legal investigation.
7. Sanctions are proscribed for destruction of discoverable evidence: court-imposed sanctions and corporation-imposed discipline.

Explain the legal concept of privilege regarding

1. Work product:
   a. Explain what it is;
   b. Describe the types of information it covers and does not cover;
   c. Specify who can assert or waive the privilege;
   d. Explain how privilege can be lost.
2. Attorney-client privilege:
   a. Explain what it is;
   b. Describe what conversations and documents are privileged;
   c. Specify who can assert or waive the privilege;
   d. Explain how the privilege can be waived or lost;
   e. Specify who is the client and under what circumstances.

Explain the need for a records management program. Provide an overview of the corporate records management program, covering

1. The necessity to keep records to satisfy legal requirements of various regulatory agencies or statutes;

2. Record retention schedules;
3. Archiving;
4. Indexing;
5. Storage, safeguards, and retrieval;
6. Periodic, routine elimination.

Sample E-Mail Evidence Policy

This policy of the corporation is to cover the entire external and internal corporate e-mail system, including microcomputers, terminals, and networks; messages, drafts, records, documents, and other information on the e-mail system including backup media and storage.

The sole purpose of the corporation's e-mail system is to assist in conducting the business of the enterprise.

All computers and communications equipment and facilities, including e-mail, and the data and information stored on them, are and remain at all times business property of the corporation and are to be used for business purposes only.

The corporation devises and maintains the security of its computing and communications systems as well as the monitoring of such systems, including e-mail. The use of all passwords and other identification and verification security devices must be made known to the corporation.

The corporation reserves the right to monitor all e-mail message content.

Any views expressed by individual employees in e-mail messages are not necessarily those of the corporation.

The corporation shall establish a systematic e-mail message, records, and document retention and destruction program designed to be effective in meeting the legitimate business needs and legal obligations of the corporation.

Violation of corporate policies by employees will invoke disciplinary measures up to and including termination of employment.

This policy will be reviewed periodically and updated in light of new legal developments and corporate experiences.

12

Creating and Managing a Web Site
Intellectual Property Considerations

Policy and Awareness

Setting up a Web site and designing a Web page for your organization can involve the use of programmers and graphic or audio-visual designers. Often this work is outsourced.

The individuals working on the Web site and page may not be familiar with the mix and requirements of intellectual property laws. It is a safe idea to assume these individuals do not know key elements of copyright and trademark law. Going on this assumption, the following is a brief review of copyright and trademark basics, plus reminders of common fallacies held regarding specific areas of the laws.

Copyright

Copyright gives owners exclusive rights to their work (a work of authorship fixed in a tangible medium) and they may authorize or license others to

- Reproduce copies of their work;
- Distribute copies of their work;
- Allow others to display or perform their work publicly.

Copyrighted works of authorship can include literary, musical, graphic, audio-visual, dramatic, and choreographic works and sound recordings. Literary works can be on paper or in an electronic medium, such as a computer program. A tangible medium can include paper, film, magnetic tape, records, and compact discs. The Internet and Web sites also are "tangible mediums of expression."

It is an infringement of the copyright for any person or organization to use, display, or distribute a copyrighted work without the owner's permission. It is also against the law to remove a copyright notice.

Public Domain Materials

Copyrightable works are in the public domain if they are created by and for the federal government, if the original copyright has expired, or if the copyright has been abandoned by the owner.

Fair Use

Our copyright system is based on the dual interests of property rights and intellectual promotion. Fair use makes the copyright law a flexible rather than rigid doctrine and a law that does not impede or stifle creative activity.

Congress codified the doctrine of fair use in the Copyright Act of 1976; however, it did not define the term, leaving its interpretation to the courts. Section 107 of the copyright act gives four factors that courts may consider in deciding whether a particular use is fair:

1. The purpose and character of the use, including whether such use is of a commercial nature or is for nonprofit educational purposes;

2. The nature of the copyrighted work;

3. The amount and substantiality of the portion used in relation to the copyrighted work as a whole; and

4. The effect of the use upon the potential market for or the value of the copyrighted work.

Congress has expressed that these four factors are neither all-inclusive nor determinative but can provide "some gauge for balancing equities." These factors, therefore, are a flexible set of guidelines for the courts to use in analyzing and deciding each individual copyright infringement case where the issue is one of fair use.

Courts must evaluate whether the use of copyrighted material was of a commercial nature or for a nonprofit educational, scientific, or historical purpose; whether the nature of the copyrighted work was published in a tangible form or was unpublished material; what amount, in quantity, and substantiality—its core or essence—of the work was used; and whether the defendant's alleged conduct had an adverse effect on the potential market or value of the copyrighted work.

Criminal and Civil Penalties

The 1992 amendments to the copyright act stiffened criminal penalties: convicted first offenders may get prison sentences for up to 5 years, fines of up to $250,000 for individuals, $500,000 for an organization; with a previous conviction, the maximum prison sentence could be 10 years.

A forfeiture clause allows the courts to seize and destroy infringing items plus "all implements, devices or equipment used in the manufacture of . . . infringing copies."

"Stealing the Best Ideas"

In designing Web sites, it is often heard that it is all right to steal ideas that work. Even though this may not lead to a copyright infringement, it could result in a charge of plagiarism. *Plagiarism* is unfair use, defined in law as appropriating the "literary composition" (book, article, advertising copy and design, etc.) or the ideas or language and "passing them off as the product of one's own mind."

The Federal Trademark Statute

The Lanham Trademark Act of 1946 (15 U.S.C., Sec. 1051-1127, 1988 ed. and Supp. V) gives a seller or producer the exclusive right to register a trademark and prevent competitors from using that trademark. Registration is in effect for 10 years; an affidavit of continued use must be filed in the sixth year. Renewal can occur any number of successive 10-year terms so long as the mark is still in commercial use. The trademark must be used in interstate commerce, although federal protection may still apply if the trademark has an effect on interstate commerce.

A producer may apply for a trademark if there is an "intent to use" the mark; the use must occur in six months or an extension can be requested.

The Lanham Act says that trademarks "include any word, name, symbol, or device, or any combination thereof." A symbol or device may be almost anything that can carry meaning.

Trade Dress

Until recently, trade dress protection basically meant original packaging; now it refers to a product's total image. Competitors therefore can be stopped from imitating the general appearance or image of a product or service if the trade dress was either inherently distinctive or had acquired a secondary meaning.

Civil Actions and Sanctions for Trademark Violations

A court-ordered temporary restraining order or an injunction is designed to protect the property of the plaintiff and could be issued to restrain ongoing or future infringing acts. The court must first decide if the party seeking to restrain was more diligent in protecting its property rights than the accused infringer.

In addition to an injunction, the court may grant an order for the seizure of goods and counterfeit marks and the means of making such marks and the destruction of all materials and equipment.

In addition to any damages sustained by the plaintiff, the court may order the defendant's profits be given the plaintiff as well as that the defendant pay the plaintiff's legal costs.

An injunction can be issued against an organization or its agent, officers, or employees.

Points to Counter Fallacies

- Unintentional copyright infringement is still infringement.
- If a work does not display a copyright notice, that does not mean the work is not copyrighted.
- Display of a work on the Internet does not mean the author has abandoned it.

- If you use copyrighted material, get permissions at each stage of the project.
- Not charging for a copy of a copyrighted work is still a violation of copyright law.
- Do not use any name or likeness of a celebrity without his or her permission; use without permission is a violation of their "right of personality."
- Photos or pictures of paintings and other artwork may be in public domain, but the gallery, museum, or archive may own the right of access; you may need permission or a license to use the photo or a copy of the artwork.
- A Web page that looks too much like another could be in violation of the Lanham Act's trade dress concept.
- A Web page design could be open to a charge of plagiarism if the copy, design elements, and layout are similar to another's Web page.
- Unauthorized promotion of another company's product or trademarked logo could violate trademark law.
- Ignorance of another's trademark is no defense to infringement.
- A new trademark must be diligently selected; that is, there should be a trademark search for identical or confusingly similar marks.
- Your trademark should have state and federal registration for either use or intent to use.

Mouse Type

Your Web page should have mouse type, which explains the ownership of intellectual property. Mouse type usually appears at the bottom of a page and in small type (4 to 6 point). You should include

- Ownership of the Web page: "[Company name]." "All rights reserved."
- Copyright notice: "Copyright © [dates] by [owner]."
- Exceptions to the copy on the Web page, such as disclaimers and limitations.
- A list of company products that appear on the page; use "Registered trademarks of [company names]."
- The first appearance of named products should have the appropriate trademark symbols ^{TM}, ^{SM}, or an ®.
- Another company's product should have "[product name] is a trademark of [company name]" if there is an agreement to do so.

Defamation and False Advertising Liability Risks

Wrongful or negligent disclosure of private or embarrassing facts usually requires such information to be communicated to more than one
person. Any disclosure of false information could lead to a defamation
suit. Defamation has two types of communication: defamation via
print, writing, pictures, or signs, called *libel*, and *slander*, which is defamation by speech. Both are communication of false information to a
third party that injures the reputation of a person or a business—causing bad opinion, public hatred, ridicule, or disgrace.

Other elements of defamation include the reasonable identification of the defamed person and damage to that person's reputation; if
the defamation refers to a public figure or is a matter of public concern,
it must be proven by the plaintiff that the defamatory language was
false and that it was communicated knowingly or with a reckless disregard as to the truth or falsity of the information. The element of falsity,
in speech of public concern cases, requires the plaintiff to prove it by
either the standard of preponderance of the evidence or, most often, the
more difficult clear and convincing evidence.

The basic defenses to defamation are that the facts of the statement are provably true and that a privilege can be invoked. Privilege
can be absolute, which is reserved for government officials, such as
judges and legislators, and the content of most public records. The press
has a qualified or limited privilege to report on matters of public interest that might go unreported. This qualified privilege can be lost if the
information is in error and malice can be shown.

Under the common law of defamation, the defendant has the burden to prove truth as a defense. In a civil case, the standard of proof normally is a preponderance of the evidence, requiring the trier of fact or
jury to believe that the existence of a fact is more probable than its nonexistence.

Commercial Defamation

For commercial defamation, the statement or representation (commercial speech) must have a tangible harm on the business, be made to a
"nonprivileged" third party, and be made by a party who was negligent
in determining if the statement was false and defamatory.

False statements by a competitor about another's products in ad-
vertising or promotion may come under the Lanham Act (15 U.S.C.,
Sect. 43(a)). In this section of the act, the Federal Trade Commission
(FTC) is charged with preventing and punishing unfair methods of com-

petition or unfair and deceptive acts or practices in or affecting commerce. The act covers false and misleading statements made about commercial products, via ads or promotions, that are believed likely to cause damage to the plaintiff's business. False disparagement of a competitor or competitor's goods is an unfair method of competition.

The FTC has the power to prohibit unfair practices and does so by first issuing a complaint of its charges to the defendant and ordering a hearing. The defendant can show cause why a cease and desist order should not be issued by the FTC. After testimony, such an order may be issued. Violation of the order by failing to stop the unfair practice could lead to a civil penalty of $10,000 for each violation. Cease and desist orders and liability can run to stockholders, directors and officers, employees, and agents.

False Advertising

Section 52(a) of the Lanham Act covers the dissemination of a false advertisement or unfair or deceptive act or practice, by any means, in or affecting commerce, that is, the purchase of goods or services. A Web site set up for promoting an organization's goods or services would surely be dissemination of commercial information.

The FTC has the power to enjoin anyone engaged in or about to disseminate false advertising. *False advertising* is defined as any ad, other than labeling, "which is misleading in a material respect." The FTC also can issue a temporary restraining order or preliminary injunction and then bring suit in a federal district court. Penalties for false advertising depend on whether the ad could injure health or if the violation is with the intent to defraud or mislead. Fines run up to $5,000 or six months in prison or both; a second conviction raises the fine to $10,000 or one year in prison or both.

Who Is a Publisher?

A critical problem for bulletin board operators, Web site owners or providers, and e-mail system owners is the definition of *publisher* or *republisher* instead of distributor or common carrier. A defamatory statement is "published" or disclosed to a third party directly or indirectly. With e-mail, a bulletin board, or a Web site, "broadcast" might be a better term. In *Stratton Oakmont Inc.* v. *Prodigy Services Co.*, Prodigy was held liable for libelous statements posted by a subscriber. The argument was over whether Prodigy was a publisher/republisher or a mere conduit of the defamatory statement. The trial judge ruled that Prodigy is a publisher with the responsibilities that go with the content it publishes; the on-line service should not be considered simply a "common carrier" with lesser responsibilities for content. The judge also pointed out that Prodigy had issued content guidelines, that Money Talk (the bulletin board where the statement was published) had an editor who was to delete messages that violated guidelines, that it used a software screening program to weed out offensive language, and that it had a bulletin board standards group. The judge said that it was Prodigy's "own policies, technology and staffing decisions which have altered the scenario and mandated a finding that it is a publisher . . . Prodigy held itself out to the public and its members as controlling the content of its computer bulletin boards."

Publication and republication are separate tortious acts: "Every repetition of the defamation is a publication in itself, even though the repeater states the source . . . The courts have said many times that the last utterance may do no less harm than the first, and that the wrong of another cannot serve as an excuse to the defendant. Likewise everyone who takes part in the publication, as in the case of the owner, editor, printer, vendor, or even carrier of a newspaper is charged with publication, although so far as strict liability is concerned the responsibility of some of these has been somewhat relaxed" (Keeton et al., 1984, Section 111 at 771).

Who is charged with publication or as being a publisher hinges on the amount of editorial control exercised over the communication. Editorial control means an element of knowledge, knowledge that the material was defamatory and therefore should not be published.

But how can true editorial control be exercised in a situation where a high volume of messages is being speedily disseminated? Public bulletin boards, Web sites, and e-mail service providers argue that there is no way they can police all this digital traffic. Corporate Web sites and e-mail systems will have a harder argument because of lower volume and more control, in theory, over how their systems are used.

For corporations, broad disclaimers on message content are one way to lower potential liability. Someone has to have the responsibility of "reasonable care" for policing a system, for taking action when a problem such as defamation or copyright infringement occurs. Either service providers or some form of self-policing organization will have to

- Handle reports of alleged abuses;
- Develop software that will identify postings and trace the source;
- Immediately investigate the allegation;

- Take action, such as deleting the offensive message;
- Provide software that screens for defamatory messages or words prior to dissemination over the network;
- Settle disputes, such as through an alternative dispute resolution method;
- Publish retractions.

Web Site Security Checklist

	YES	NO
Is the security of your Internet connections adequate?	☐	☐
Is there access control technology or a complete firewall system for the Web server and other computing and communications equipment with on-line connections?	☐	☐
Does the Web site provider have a comprehensive security policy that is communicated to its employees?	☐	☐
Is this policy enforced?	☐	☐
Are the employees trained to be competent in security procedures and technology?	☐	☐
Are all security measures monitored and audited?	☐	☐
Can you answer "yes" to the preceding questions for your operation?	☐	☐
What will the Internet or Web site provider do if service is interrupted?	☐	☐

13

Computer Crime Law

Legal remedies to the wide and ever-growing types of computer-related crime are to be found in federal and state statutes. Prosecutors may choose to use either computer crime statutes or related criminal statutes that might be "shoehorned" to fit a specific criminal act.

The following chapters and the appendices to this chapter present laws that allow an organization to take action against those that breach information system security or harm information assets. These laws include federal and state statutes on computer and telecommunications crime plus related statutes on false statements and claims, theft, fraud, embezzlement, transport of stolen goods, malicious destruction of property, conspiracy, and forfeiture. Each statute has an analysis of its provisions, key phrases and terms, and leading cases prosecuted under the statute. The Federal Sentencing Guidelines for individuals and organizations convicted under specific federal statutes are discussed. Also included is an overview of state laws covering computer-related crimes.

Federal Statutes on Computer-Related Crime

The Computer Fraud and Abuse Act of 1986 has become the key federal statute for information security and prosecution of computer-related crime. Expanded liability for computer-related fraud and provisions for civil remedies are provided in the 1994 amendments to the act. The amendments also provide a lower standard of liability for those who knowingly gain access to a computer without authorization and those who commit acts with a "reckless disregard" of a substantial and unjustifiable risk to a computer or computer system.

The National Information Infrastructure Protection Act of 1996 became Public Law 104-294 in October 1996. The act amends the Computer Fraud and Abuse Act (Title 18 U.S.C., Sect. 1030). The amendments increase the protection of the national information infrastructure, criminalize access by government employees who exceed their authority to access government information, and extend protection to computers and communication used in interstate or foreign commerce.

New subsections ((e)(8) through (e)(8)(D)) to the code include a definition of *damage*, meaning "any impairment to the integrity or availability of data, a program, a system, or information, that: causes loss aggregating at least $5,000 in value during any one-year period to one or more individuals; modifies or impairs, or potentially modifies or impairs, the medical examination, diagnosis, treatment or care of one or more individuals; causes physical injury to any person; or threatens public health or safety."

The amendments criminalize access by government employees who exceed or abuse their authority to gain access to and obtain government and private sector information and criminalize extortion by a person or firm that transmits "any communication in interstate or foreign commerce containing any threat to cause damage to a protected computer." *Obtaining information* includes simply reading the information, as well as copying or stealing it.

The amendments change the punishment for offenses committed for purposes of commercial advantage or private financial gain or in furtherance of any criminal or tortious act in violation of federal or state law. These offenses carry a fine or imprisonment of not more than five years or both.

In the computer definition section, *federal interest* is dropped for *protected* computer, to include government or financial institution computers and any computer used in interstate or foreign commerce or communication.

The complete text of the Computer Fraud and Abuse Act is provided in Appendix A of this chapter.

The Electronic Communications Privacy Act (PL 99-508, 18 U.S.C., Sect. 2510-2520 and 2701-2710) amends the federal electronic surveillance statutes originally enacted as Title III of the Omnibus Crime Control and Safe Streets Act of 1968. The ECPA extends the protection of the Wiretap Act of 1968 to electronic communications and communications systems, including radio, satellite, and data communications. Excluded from coverage is any radio transmission "readily accessible to the general public" and types of protected radio signals.

The ECPA protects the privacy of transmitted and stored electronic communications from interception and disclosure. The act covers electronic communications made by "any transfer of signs, signals, writings, images, sounds, data or intelligence of any nature transmitted in whole or in part by a wire, radio, electromagnetic, photoelectronic or photo-optical system that affects interstate or foreign commerce." The only requirement of the act is that the information affect interstate or foreign commerce. The act applies to government, private, and public systems, exempting those systems without any expectation of privacy.

Under the ECPA, it is illegal to intercept electronic communications or to use or disclose the contents to another person. Further, it is a felony carrying fines and prison terms. The party whose communication was intercepted can bring suit in federal court. Penalties are greater when the interception was for commercial advantage or illegal purpose, and the plaintiff may recover actual damages and profits made as a result of the violation.

The privacy of electronic mail is protected under the law, with a misdemeanor penalty for those who break into an electronic communications system holding messages. The ECPA covers any service "which provides to users thereof the ability to send or receive wire or electronic communications." A remote computing service is a system that provides public computer communications storage or processing. Also covered is "any person or entity providing the wire or electronic communication service." A key and implied element is that the system be configured for privacy.

Offenses committed for commercial advantage or malicious destruction or damage carry a fine of up to $250,000, a one-year prison term, or both.

For both stored and transmitted communications, the intent standard as described in this computer crime law applies to any defendant; that is, it must be proven that the defendant intentionally sought to intercept, alter, damage, or destroy the data communication or message.

For providers of electronic communications services to the public it is illegal to divulge knowingly the contents of any communication except to the sender or intended recipient of the information.

Wiretap Authority

Under the ECPA, law enforcement agencies must obtain the approval of certain high-level Justice Department officials, then a court order authorizing or approving their proposed interception.

Let us turn to the specifics of the review process. The provisions of Title III specifically assign the review powers to the attorney general, but allow this authority to be delegated to other Department of Justice officials. The Department of Justice (DOJ) review process "must occur prior to the submission to the court of an application for interception. Such review and approval must, in almost all instances, precede the actual interception. However, in certain 'emergency' situations, interception may temporarily precede application to the court. In those instances, the Department's authorization must still be obtained prior to interception, and the application to the court must be submitted within 48 hours of the interception." Applications for electronic surveillance are reviewed initially by the Electronic Surveillance Unit of the Criminal Division's Office of Enforcement Operations.

A key restriction on the use of electronic surveillance by law enforcement is the requirement that the government obtain an order from a court of competent jurisdiction prior to the use of most types of electronic surveillance. A court of competent jurisdiction is defined at 18 U.S.C. 3127(2) as "(A) a district court of the United States (including a magistrate [United States magistrate judge] of such a court) or a United States Court of Appeals; or (B) a court of general criminal jurisdiction of a State authorized by the law of that State to enter orders authorizing the use of pen register or a trap and trace device . . . "

As noted already, the ECPA broadened the definition of communications to include electronic communications, including computers, fax machines, and paging devices.

The application must be specific enough for the court to conclude

1. That probable cause exists that certain listed persons have committed, are committing, or will commit offenses that are proper predicates for the specific type of electronic surveillance;

2. That probable cause exists that all or some of these persons have used, are using, or will use a targeted facility or targeted premises in connection with the commission of predicate offenses (under Title III or ECPA); and

3. That probable cause exists that the targeted facility has been used, is being used, or will be used in connection with the predicate offenses.

The court application also must contain "a complete statement as to other investigative procedures that have been tried and failed, or reasonably appear unlikely to succeed if tried, or which would be too dangerous to employ, and a complete statement of all other applications for electronic surveillance involving the persons, facilities, or premises which are subject to the current application."

The ECPA authorizes applications for the "roving interception" of wire communications. "An application for the interception of wire communications without specifying the facility or facilities to be targeted may be made in those instances where it can be shown that the subject or subjects of the interception have demonstrated a purpose to thwart interception by changing facilities."

The ECPA contains a provision that authorizes an emergency interception of electronic communications before a court authorization can be obtained where "such officer is specifically designated, prior to the interception, by the Attorney General, Deputy Attorney General, or Associate Attorney General and where an emergency situation exists that involves (1) immediate danger of death or serious bodily injury to any person, (2) conspiratorial activities threatening the national security interest, or (3) conspiratorial activities characteristic of organized crime. The statute requires that grounds must exist under which an order could be entered to authorize the interception and that an application be made within 48 hours after the interception has occurred or begins to occur. If a court order is obtained within that time frame, the interception may continue as ordered."

If the application is denied, the contents of the interception must be treated as a violation of Title III, and an inventory of the interception served on the party named in the application.

Stored Electronic Communications

ECPA protection extends to electronic communications stored after transmission. Requirements for government access are set forth in 18 U.S.C. 2703-2705: a search warrant issued under the Federal Rules of Criminal Procedure (Rule 41, search and seizure) or an equivalent state warrant. The government also can use an administrative subpoena authorized by a federal or state statute or a federal or state grand jury or trial subpoena. Another avenue is via a court order for disclosure if the government can show "there is reason to believe the contents of a wire or electronic communication, or the records or other information sought, are relevant to a legitimate law enforcement inquiry."

Court Decision on Stored Electronic Communications

In the case *Steve Jackson Games* (SJG) v. *United States Secret Service*, SJG filed a civil suit in federal court against the Secret Service under Title I of the Electronic Communications Privacy Act (18 U.S.C., Sect. 2511(1)(a)). A federal district court had found that the Secret Service violated Title II of the ECPA by seizing stored electronic communications (e-mail) in SJG's computers without complying with statutory provisions. The district court did not find that the Secret Service "intercepted" the e-mail in violation of Title I.

The issue that came before the U.S. Court of Appeals for the Fifth Circuit was the definition of *interception*; specifically, the seizure of a computer on which is stored private e-mail that has been sent to an electronic bulletin board but not yet read by the recipients. Section 2511(1)(a) proscribes "intentionally intercepting . . . any wire, oral, or electronic communications" unless authorized by court order. Section 2520 allows persons whose electronic communications are intercepted to bring a civil suit for damages.

In the ECPA, *intercept* is defined as "the aural or other acquisition of the contents of any wire, electronic, or oral communication through the use of any electronic, mechanical, or other device."

SJG argued that the information or e-mail could still be intercepted under the ECPA even though it was not in transit and argued that the intent of Congress was to protect e-mail and bulletin boards.

The appeals court said the

> language of the Act [Title I] controls . . . electronic storage is defined as any temporary, intermediate storage of a wire or electronic communication incidental to the electronic transmission thereof . . . The E-mail in issue was in "electronic storage." Congress' use of the word "transfer" in the definition of "electronic communication," and its omission in that definition of the phrase "any electronic storage of such communication" reflects that Congress did not intend for "intercept" to apply to electronic communications in storage . . . We find no indication in either the act or its legislative history that Congress intended for conduct that is clearly prohibited by Title II to furnish the basis for a civil remedy under Title I as well.

Counterfeit Access Devices

When Congress passed the counterfeit access device law (18 U.S.C., Sect. 1029) on the use of access devices to carry out fraud and other crimes, it was focusing primarily on credit card fraud. However, the law was intended to be broad enough to encompass technological advances, including, for example, telephone and cellular access codes.

In *United States* v. *Fernandez*, the court held that computer passwords met the definition of an access device that, if possessed and used with the intent to defraud, could come under 18 U.S.C., Sect. 1029.

Fraudulent use of telephone access codes or credit card account numbers also is reached by the statute. The complete text of the statute is given in Appendix B to this chapter.

Drop-Dead Functions Curbed

Tucked away in the Omnibus Crime Bill is an important amendment to the Computer Fraud and Abuse Act (CFAA) (18 U.S.C., Sect. 1030). In Section 290001 of the Crime Bill is the Computer Abuse Amendments Act of 1994, which prohibits the knowing transmission of a "program, information, code, or command" to a computer or computer system that is intended to damage, deny, or interrupt a computer or a computer system if the harmful transmission occurs without the authorization of the computer system owner and results in damage of $1,000 or more in any one-year period.

Also prohibited is the transmission of codes or commands if done "with reckless disregard of a substantial and unjustifiable risk" that the transmission damage or deny the use of a computer, computer system, network, computer program or data.

Penalties include a fine, one year in prison, or both.

Legislative Intent

Senator Leahy (D-Vt.), the author of the amendments, gave a statement to the Senate clarifying the legislative intent:

> This provision clarifies the intent standards, the actions prohibited and the jurisdiction of the current Computer Fraud and Abuse Act (CFAA). Under the current statute, prosecution of computer abuse crimes must be predicated upon the violator's gaining "unauthorized access" to the "affected federal interested computers." However, computer abusers have developed an arsenal of new techniques which result in the replication and transmission of destructive programs or codes that inflict damage upon remote computers to which the violator never gained "ac-

cess" in the commonly understood sense of that term. This new section of the CFAA created by this bill places the focus on harmful intent and resultant harm, rather than on the technical concept of computer "access."

During consideration of the legislation, manufacturers of software raised the issue of whether this statute would criminalize the use of so-called disabling codes which computer software copyright owners sometimes use to enforce their license agreements " . . . it is not the intention of this legislation to criminalize the use of disabling codes when their use is pursuant to a lawful licensing agreement that specifies the conditions for reentry or software disablement."

Provisions of the Amendments

The amendments prohibit malicious transmission that "modifies or impairs, or potentially modifies or impairs, the medical examination, medical diagnosis, medical treatment, or medical care of one or more individuals . . . "

A civil action section is added, whereby any person who suffers damage from a person who knowingly causes a malicious transmission "may maintain a civil action against the violator to obtain compensatory damages and injunctive relief or other equitable relief." A malicious transmission caused by "reckless disregard" can bring only economic damages.

Actions brought under this amendment must be within two years of the date of the act complained of or the date of the discovery of the damage.

The last change was to insert *adversely* in Section 1030(a)(3) to read: "and such conduct [intentional and unauthorized access] adversely affects the use of the Government's operation of such computer."

Definitions of Terms in the Amendments

Turning to the terms used in the amendments, *a computer used in interstate commerce or communications* is a broader term than *federal interest* computer.

Knowingly causes means acting with knowledge, cognizance, and awareness of the nature of one's conduct.

Reckless disregard refers to a lack of due caution, heedless indifference; a *substantial and unjustifiable risk* would have to be defined situation by situation.

Use of the terms *program, information,* and *code or command* includes more than logic bombs and should cover worms, viruses, or any harmful device or function.

Use of Drop-Dead Functions

Drop-dead functions or disablement devices have been used by software developers to deny access or use of their software in a contract dispute or license breach or for payment default. Deactivation of software can be accomplished either through a "time or logic bomb" in the program, or by dialing into the lessee's computer and inserting a deactivation code. A logic bomb in the software, for instance, may be set to the lessee's payment schedule. Usually drop-dead functions are incorporated secretly into a software program.

Software disablement by the developer or lessor has been called a form of repossession and an unlawful denial of the lessee's use of his information in the computer. Common law repossession is codified under Article 9 of the Uniform Commercial Code and is clear on how a secured creditor can go about repossessing his property; however, in the case of computer software, it is not clear how software can be "repossessed" without violating trespass and seizure of the debtor's property, as well as the principles of fairness.

In *Franks & Sons, Inc.* v. *Information Solutions, Inc.,* Franks & Sons, Inc., a trucking company, contracted with the software developer Information Solutions for the sale of a computer hardware and software system with a maintenance agreement plus a licensing agreement for proprietary software designed for the trucking business. Franks withheld full payment due to bugs in the software that made the system not fully operational. Information Solutions then told Franks the software had a drop-dead function and threatened to use it. Franks then sought an injunction against the use of the disablement device and damages from the harm that would result from a shutdown of Franks's business.

The court said enforcement of a computer software purchase contract was contrary to public policy where the software developer had included in its product a surprise "drop-dead" device, "which chills the functioning of any business whose operation is a slave to a computer." The drop-dead device was not contained in the original agreement; the court said this surely would have affected the original contract negotiations.

Avoiding Abhorrent Contracts

Every software contract should reflect a balance of interests. The software developer is concerned with getting paid for its product and protection against software piracy. The licensee wants software that functions according to the promised specifications.

The amendments to the CFAA and the *Franks* case strongly suggest a review of software licenses and the use of disabling functions. Disabling codes can be used if they are part of a law licensing agreement or contract that specifies the conditions of use. A review of agreements and procedures, at a minimum, should cover

1. Revealing the existence of a device or technique for disabling the software program;

2. Adequate prior notice of any disabling action;

3. Consent of the computer system owner to conditions under which entry to its computer system or deactivation could occur;

4. Conditions under which software could be repossessed.

Appendix A: Computer Fraud and Abuse Act of 1986 [Revised text, 1996], PL 99-474 (18 U.S.C., Sect. 1030)

Section 1030 covers fraud and related activity in connection with computers:

> (a) Whoever—
> > (1) knowingly having accessed a computer without authorization or exceeding authorized access, and by means of such conduct having obtained information that has been determined by the United States Government pursuant to an Executive order or statute to require protection against unauthorized disclosure for reasons of national defense or foreign relations, or any restricted data, as defined in paragraph [(y)] of section 11 of the Atomic Energy Act of 1954 [42 U.S.C.S., Sect. 2014(y)], with reason to believe that such information so obtained could be used to the injury of the United States, or to the advantage or any foreign nation willfully communicates, delivers, transmits, or causes to be communicated, delivered, or transmitted, or attempts to communicate, deliver, or cause to be communicated, delivered, or transmitted the same to any per-

son not entitled to receive it, or willfully retains the same and fails to deliver it to the officer or employee of the United States entitled to receive it;

(2) intentionally accesses a computer without authorization or exceeds authorized access, and thereby obtains—

(A) information contained in a financial record of a financial institution, or of a card issuer as defined in section 1602(n) of title 15, or contained in a file of a consumer reporting agency on a consumer, as such terms are defined in the Fair Credit Reporting Act (15 U.S.C. 1681 et seq.);

(B) information from any department or agency of the United States; or

(C) information from any protected computer if the conduct involved an interstate or foreign communication;

(3) intentionally, without authorization to access any nonpublic computer of a department or agency of the United States, accesses such a computer of that department or agency that is exclusively for the use of the Government of the United States or, in the case of a computer not exclusively for such use, is used by or for the Government of the United States and such conduct affects that use by or for the Government of the United States;

(4) knowingly and with intent to defraud, accesses a protected interest computer without authorization, or exceeds authorized access, and by means of such conduct furthers the intended fraud and obtains anything of value, unless the object of the fraud and the thing obtained consists only of the use of the computer and the value of such use is not more than $5,000 in any 1-year period;

(5) (A) knowingly causes the transmission of a program, information, code, or command, and as a result of such conduct, intentionally causes damage without authorization, to a protected computer;

(B) intentionally accesses a protected computer without authorization, and as a result of such conduct, recklessly causes damage; or

(C) intentionally accesses a protected computer without authorization, and as a result of such conduct, causes damage;

(6) knowingly and with intent to defraud traffics (as defined in section 1029) in any password or similar information

through which a computer may be accessed without authorization, if—

    (A) such trafficking affects interstate or foreign commerce; or

    (B) such computer is used by or for the Government of the United States;

(7) with intent to extort from any person, firm, association, educational institution, financial institution, government entity, or other legal entity, any money or other thing of value, transmits in interstate or foreign commerce any communication containing any threat to cause damage to a protected computer;

(b) Whoever attempts to commit an offense under subsection (a) of this section shall be punished as provided in subsection (c) or this section.

(c) The punishment for an offense under subsection (a) or (b) of this section is—

    (1) (A) a fine under this title or imprisonment for not more than 10 years, or both, in the case of an offense under subsection (a)(1) of this section which does not occur after a conviction for another offense under this section, or an attempt to commit an offense punishable under this subparagraph; and

        (B) a fine under this title or imprisonment for not more than 20 years, or both, in the case of an offense under subsection (a)(1) of this section which occurs after a conviction for another offense under such subsection, or an attempt to commit an offense punishable under this subparagraph; and

    (2) (A) a fine under this title or imprisonment for not more than one year, or both, in the case of an offense under subsection (a)(2), (a)(3), (a)(5)(C) or (a)(6) of this section which does not occur after a conviction for another offense under this section, or an attempt to commit an offense punishable under this subparagraph; and

        (B) a fine under this title or imprisonment for not more than 5 years or both, in the case of an offense under subsection (a)(2), if—

            (i) the offense was committed for purposes of commercial advantage or private financial gain;

            (ii) the offense was committed in furtherance of any criminal or tortious act in violation of the Constitution or laws of the United States or of any State; or

    (iii) the value of the information obtained exceeds $5,000;

  (C) a fine under this title or imprisonment for not more than 10 years, or both, in the case of an offense under this section (a)(2), (a)(3), or (a)(6) of this section which occurs after a conviction for another offense under such subsection, or an attempt to commit an offense punishable under this subparagraph and;

(3) (A) a fine under this title or imprisonment for not more than five years, or both, in the case of an offense under subsection (a)(4), (a)(5)(A), or (a)(7) of this section which does not occur after a conviction for another offense under this section, or an attempt to commit an offense punishable under this subparagraph; and

  (B) a fine under this title or imprisonment for not more than 10 years, or both, in the case of an offense under subsection (a)(4), (a)(5)(A), (a)(5)(B), (a)(5)(C), or (a)(7) of this section which occurs after a conviction for another offense under this section, or an attempt to commit an offense punishable under this subparagraph and;

(d) The United States Secret Service shall, in addition to any other agency having such authority, have the authority to investigate offenses under subsections (a)(2)(A), (a)(2)(B), (a)(3), (a)(4), (a)(5), and (a)(6) of this section. Such authority of the United States Secret Service shall be exercised in accordance with an agreement which shall be entered into by the Secretary of the Treasury and the Attorney General.

(e) As used in this section—

  (1) the term "computer" means an electronic, magnetic, optical, electrochemical, or other high speed data processing device performing logical, arithmetic, or storage functions, and includes any data storage facility or communications facility directly related to or operating in conjunction with such device, but such term does not include an automated typewriter or typesetter, a portable hand held calculator, or other similar device;

  (2) the term "protected" means a computer—

    (A) exclusively for the use of a financial institution or the United States Government, or, in the case of a computer not exclusively for such use, used by or for a financial institution or the United States Government and the conduct constituting the offense affects that use by or for the financial institution or the Government; or

(B) which is used in interstate or foreign commerce or communication;

(3) the term "State" includes the District of Columbia, the Commonwealth of Puerto Rico, and any other possession or territory of the United States;

(4) the term "financial institution" means—

(A) a bank with deposits insured by the Federal Deposit Insurance Corporation;

(B) the Federal Reserve or a member of the Federal Reserve including any Federal Reserve Bank;

(C) an institution with accounts insured by the Federal Savings and Loan Insurance Corporation;

(D) a credit union with accounts insured by the National Credit Union Administration;

(E) a member of the Federal home loan bank system and any home loan bank;

(F) any institution of the Farm Credit System under the Farm Credit Act of 1971;

(G) a broker-dealer registered with the Securities and Exchange Commission pursuant to section 15 of the Securities Exchange Act of 1934; and

(H) the Securities Investor Protection Corporation;

(5) the term "financial record" means information derived from any record held by a financial institution pertaining to a customer's relationship with the financial institution;

(6) the term "exceeds authorized access" means to access a computer with authorization and to use such access to obtain or alter information in the computer that the accessor is not entitled so to obtain or alter;

(7) the term "department of the United States" means the legislative or judicial branch of the Government or one of the executive departments enumerated in section 101 of title 5 and;

(8) the term "damage" means any impairment to the integrity or availability of data, a program, a system, or information, that—

(A) causes loss aggregating at least $5,000 in value during any one year period to one or more individuals;

      (B) modifies or impairs, or potentially modifies or impairs, the medical examination, diagnosis, treatment, or care of one or more individuals;

      (C) causes physical injury to any person; or

      (D) threatens public health or safety; and

  (9) the term "government entity" includes the Government of the United States, any State or political subdivision of the United States, any foreign country, and any state, province, municipality, or other political subdivision of a foreign country.

(f) This section does not prohibit any lawfully authorized investigative, protective, or intelligence activity of a law enforcement agency of the United States, a State, or a political subdivision of a State, or of an intelligence agency of the United States.

(g) Any person who suffers damage or loss by reason of a violation of the section, may maintain a civil action against the violator to obtain compensatory damages and injunctive relief or other equitable relief. Damages for violations involving damages as defined in subsection (e)(8)(A) are limited to economic damages. No action may be brought under this subsection unless such action is begun within two years of the date of the act complained of or the date of the discovery of the damage.

(h) The Attorney General and the Secretary of the Treasury shall report to the Congress annually, during the first three years following the date of the enactment of this subsection, concerning investigations and prosecutions under section 1030(a)(5) of title 18, United States Code.

Appendix B: Counterfeit Access Device (18 U.S.C., Sect. 1029)

Section 1029 covers fraud and related activity in connection with access devices:

(a) Whoever—

  (1) knowingly and with intent to defraud produces, uses, or traffics in one or more counterfeit access devices;

  (2) knowingly and with intent to defraud traffics in or uses one or more unauthorized access devices during any one-year period, and by such conduct obtains anything of value aggregating $1,000 or more during that period;

(3) knowingly and with intent to defraud possesses 15 or more devices which are counterfeit or unauthorized access devices; or

(4) knowingly, and with intent to defraud, produces, traffics in, has control or custody of, or possesses device-making equipment;

(5) knowingly and with intent to defraud effects transactions, with one or more access devices issued to another person or persons, to receive payment or any other thing of value during any one-year period the aggregate value of which is equal to or greater than $1,000;

(6) without the authorization of the issuer of the access device, knowingly and with intent to defraud solicits a person for the purpose of—

(A) offering an access device; or

(B) selling information regarding or an application to obtain an access device; [or]

(7) without the authorization of the credit card system member or its agent, knowingly and with intent to defraud causes or arranges for another person to present to the member or its agent, for payment, one or more evidences or records of transactions made by an access device;

(8) knowingly and with intent to defraud uses, produces, traffics in, has control or custody of, or possesses—

(A) a scanning receiver; or

(B) hardware or software used for altering or modifying telecommunications instruments to obtain unauthorized access to telecommunications services,

shall, if the offense affects interstate or foreign commerce, be punished as provided in subsection (c) of this section.

(b) (1) Whoever attempts to commit an offense under subsection (a) of this section shall be punished as provided in subsection (c) of this section.

(2) Whoever is a party to a conspiracy of two or more persons to commit an offense under subsection (a) of this section, if any of the parties engages in any conduct in furtherance of such offense, shall be fined an amount not greater than the amount provided as the maximum fine for such offense under subsection (c) of this section or imprisonment not longer than one-half the period provided as the maximum imprisonment for such offense under subsection (c) of this section, or both.

(c) The punishment for an offense under subsection (a) or (b)(1) or this section is—

    (1) a fine of not more than the greater of $10,000 or twice the value obtained by the offense or imprisonment for not more than 10 years, or both, in the case of an offense under subsection (a)(2) or (a)(3) of this section which does not occur after a conviction for another offense under either such subsection, or an attempt to commit an offense punishable under this paragraph;

    (2) a fine of not more than the greater of $50,000 or twice the value obtained by the offense or imprisonment for not more than 15 years, or both, in the case of an offense under subsection (a)(1) or (a)(4) of this section which does not occur after a conviction for another offense under either such subsection, or an attempt to commit an offense punishable under this paragraph; and

    (3) a fine of not more than the greater of $100,000 or twice the value obtained by the offense or imprisonment for not more than 20 years, or both, in the case of an offense under subsection (a) of this section which occurs after a conviction for another offense under such subsection, or an attempt to commit an offense punishable under this paragraph.

(d) The United States Secret Service shall, in addition to any other agency having such authority, have the authority to investigate offenses under this section. Such authority of the United States Secret Service shall be exercised in accordance with an agreement which shall be entered into by the Secretary of the Treasury and the Attorney General.

(e) As used in this section—

    (1) the term "access device" means any card, plate, code, account number, or other means of account access that can be used, alone or in conjunction with another access device, to obtain money, goods, services, or any other thing of value, or that can be used to initiate a transfer of funds (other than a transfer originated solely by paper instrument);

    (2) the term "counterfeit access device" means any access device that is counterfeit, fictitious, altered, or forged, or an identifiable component of an access device or a counterfeit access device;

    (3) the term "unauthorized access device" means any access device that is lost, stolen, expired, revoked, canceled, or obtained with intent to defraud;

(4) the term "produce" includes design, alter, authenticate, duplicate, or assemble;

(5) the term "traffic" means transfer, or otherwise dispose of, to another, or obtain control of with intent to transfer or dispose of; and

(6) the term "device-making equipment" means any equipment, mechanism, or impression designed or primarily used for making an access device or a counterfeit access device.

(f) This section does not prohibit any lawfully authorized investigative, protective, or intelligence activity of a law enforcement agency of the United States, a State, or a political subdivision of a State, or of an intelligence agency of the United States, or any activity authorized under chapter 224 of this title. For purposes of this subsection, the term "State" includes a State of the United States, the District of Columbia, and any commonwealth, territory, or possession of the United States.

14

Federal Sentencing Guidelines for Individuals and Organizations

Sentencing Guidelines for Individuals Charged with Computer-Related Fraud

The sentencing guidelines for federal crimes have been prepared by the United States Sentencing Commission. The commission, established by the Sentencing Reform Act of 1984, was charged with drafting guidelines for federal judges to use when sentencing convicted defendants. The objective of the act was an effective, fair sentencing system. Congress sought to obtain honesty, reasonable uniformity, and proportionality in sentencing. Honesty means the sentence imposed by the court is the sentence the offender will serve; abolishing parole was one method to this end. Narrowing the wide disparity in sentences imposed for similar criminal offenses committed by similar offenders is a way to achieve reasonable uniformity. And imposing appropriately different sentences for criminal conduct of differing severity could achieve the objective of proportionality in sentencing.

The act directs the commission to create categories of offense behavior and offender characteristics. In 1986, the commission asked for and got legislation to deal with issuance of general policy statements concerning imposition of fines, the permissible width of a guideline range calling for a term of imprisonment, and appellate review of sentences.

A federal court must select a sentence from within the guideline range. In an atypical case, the court is allowed a departure from the guideline but must specify reasons for the departure; such a departure may be reviewed by an appellate court.

The federal sentencing guidelines took effect on November 1, 1987.

The Sentencing Table and Instructions

The commission established a sentencing table (Table 14-1) that, for technical and practical reasons, contains 43 levels. Each level in the table prescribes ranges that overlap with the ranges in the preceding and succeeding levels.

In applying the guidelines, one must first determine the applicable offense guideline section. For example, for computer-related fraud, 18 U.S.C. 1029 and 1030, this would be Part F, Offenses Involving Fraud or Deceit, 2F1.1.

Next determine the base offense level and apply any appropriate specific offense characteristics contained in the particular guideline. Each offense may cover one statute or many, has a corresponding base offense level, and may have one or more specific characteristics that adjust the offense level upward or downward. Other instructions must be applied depending on the particular offense and case.

The following discussion presents the guidelines for fraud, including computer-related fraud, with comments on sentencing-related cases and instructions. This guideline is designed to apply to a wide variety of fraud cases. The statutory maximum term of imprisonment for most such offenses is five years. The guideline does not link the characteristics of an offense to specific code sections. Because federal fraud statutes are written so broadly, a single pattern of offense conduct usually can be prosecuted under several code sections, as a result of which the offense of conviction may be somewhat arbitrary. Furthermore, most fraud statutes cover a broad range of conduct with extreme variation in severity.

Empirical analyses of preguideline practices showed that the most important factors that determined sentence length were the amount of loss and whether the offense was an isolated crime of opportunity or was sophisticated or repeated. Accordingly, although they are imper-

Table 14-1 Sentencing Table (in months of imprisonment).
Source: U.S. Sentencing Commission Guidelines, November 1, 1994.

	Offense Level	Criminal History Category (Criminal History Points)					
		I (0 or 1)	II (2 or 3)	III (4, 5, 6)	IV (7, 8, 9)	V (10, 11, 12)	VI (13 or more)
Zone A	1	0–6	0–6	0–6	0–6	0–6	0–6
	2	0–6	0–6	0–6	0–6	0–6	1–7
	3	0–6	0–6	0–6	0–6	2–8	3–9
	4	0–6	0–6	0–6	2–8	4–10	6–12
	5	0–6	0–6	1–7	4–10	6–12	9–15
	6	0–6	1–7	2–8	6–12	9–15	12–18
	7	0–6	2–8	4–10	8–14	12–18	15–21
	8	0–6	4–10	6–12	10–16	15–21	18–24
	9	4–10	6–12	8–14	12–18	18–24	21–27
Zone B	10	6–12	8–14	10–16	15–21	21–27	24–30
Zone C	11	8–14	10–16	12–18	18–24	24–30	27–33
	12	10–16	12–18	15–21	21–27	27–33	30–37
	13	12–18	15–21	18–24	24–30	30–37	33–41
	14	15–21	18–24	21–27	27–33	33–41	37–46
	15	18–24	21–27	24–30	30–37	37–46	41–51
	16	21–27	24–30	27–33	33–41	41–51	46–57
	17	24–30	27–33	30–37	37–46	46–57	51–63
	18	27–33	30–37	33–41	41–51	51–63	57–71
	19	30–37	33–41	37–46	46–57	57–71	63–78
	20	33–41	37–46	41–51	51–63	63–78	70–87
	21	37–46	41–51	46–57	57–71	70–87	77–96
	22	41–51	46–57	51–63	63–78	77–96	84–105
	23	46–57	51–63	57–71	70–87	84–105	92–115
	24	51–63	57–71	63–78	77–96	92–115	100–125
Zone D	25	57–71	63–78	70–87	84–105	100–125	110–137
	26	63–78	70–87	78–97	92–115	110–137	120–150
	27	70–87	78–97	87–108	100–125	120–150	130–162
	28	78–97	87–108	97–121	110–137	130–162	140–175
	29	87–108	97–121	108–135	121–151	140–175	151–188
	30	97–121	108–135	121–151	135–168	151–188	168–210
	31	108–135	121–151	135–168	151–188	168–210	188–235
	32	121–151	135–168	151–188	168–210	188–235	210–262
	33	135–168	151–188	168–210	188–235	210–262	235–293
	34	151–188	168–210	188–235	210–262	235–293	262–327
	35	168–210	188–235	210–262	235–293	262–327	292–365
	36	188–235	210–262	235–293	262–327	292–365	324–405
	37	210–262	235–293	262–327	292–365	324–405	360–life
	38	235–293	262–327	292–365	324–405	360–life	360–life
	39	262–327	292–365	324–405	360–life	360–life	360–life
	40	292–365	324–405	360–life	360–life	360–life	360–life
	41	324–405	360–life	360–life	360–life	360–life	360–life
	42	360–life	360–life	360–life	360–life	360–life	360–life
	43	life	life	life	life	life	life

fect, these are the primary factors upon which the guideline has been based.

The extent to which an offense is planned or sophisticated is important in assessing its potential harmfulness and the dangerousness of the offender, independent of actual harm. A complex scheme or repeated incidents of fraud are indicative of an intention and potential to do considerable harm. In preguideline practice, this factor had a significant impact, especially in frauds involving small losses. Accordingly, the guideline specifies a two-level enhancement when this factor is present.

A defendant who has been subject to civil or administrative proceedings for the same or similar fraudulent conduct demonstrates aggravated criminal intent and is deserving of additional punishment for not conforming with the requirements of judicial process or orders issued by federal, state, or local administrative agencies.

Offenses that involve the use of transactions or accounts outside the United States in an effort to conceal illicit profits and criminal conduct involve a particularly high level of sophistication and complexity. These offenses are difficult to detect and require costly investigations and prosecutions. Consequently, a minimum level of 12 is provided for these offenses.

Sentencing Guidelines for Individuals

These offenses, involving computer-related fraud, have a base level of 6.

Specific Offense Characteristics

If the loss exceeded $2,000, increase the offense level as follows:

Loss (Apply the Greatest)	Increase in Level
A. $2,000 or less	No increase
B. More than $2,000	Add 1
C. More than $5,000	Add 2
D. More than $10,000	Add 3
E. More than $20,000	Add 4
F. More than $40,000	Add 5
G. More than $70,000	Add 6
H. More than $120,000	Add 7
I. More than $200,000	Add 8

J.	More than $350,000	Add 9
K.	More than $500,000	Add 10
L.	More than $800,000	Add 11
M.	More than $1.5 million	Add 12
N.	More than $2.5 million	Add 13
O.	More than $5 million	Add 14
P.	More than $10 million	Add 15
Q.	More than $20 million	Add 16
R.	More than $40 million	Add 17
S.	More than $80 million	Add 18

If the offense involved more than minimal planning or a scheme to defraud more than one victim, the offense increases by two levels.

If the offense involved a misrepresentation that the defendant was acting on behalf of a charitable, educational, religious, or political organization or a government agency, or the violation of any judicial or administrative order, injunction, decree, or process not addressed elsewhere in the guidelines, it increases by two levels. If the resulting offense level is less than level 10, it increases to level 10.

If the offense involved the conscious or reckless risk of serious bodily injury, it increases by two levels. If the resulting offense level is less than level 13, it increases to level 13.

If the offense involved the use of foreign bank accounts or transactions to conceal the true nature or extent of the fraudulent conduct and the offense level as determined by the preceding text is less than level 12, it increases to level 12.

If the offense substantially jeopardized the safety and soundness of a financial institution or affected a financial institution and the defendant derived more than $1 million in gross receipts from the offense, it increases by four levels. If the resulting offense level is less than level 24, it increases to level 24.

Application Notes

More than minimal planning (subsection (b)(2)(a)) means more planning than is typical for commission of the offense in a simple form. This condition also exists if significant affirmative steps were taken to conceal the offense. "More than minimal planning" is deemed present in any case involving repeated acts over a period of time, unless it is clear that each instance was purely opportune.

Scheme to defraud more than one victim (subsection (b)(2)(B)) refers to a design or plan to obtain something of value from more than one person. In this context, *victim* refers to the person or entity from which the funds are to come directly.

The base offense level for 18 U.S.C., Section 1030(a)(1), "knowingly accesses a computer . . . and . . . obtains information . . . ," is set at 35 if the information is top secret and at 30 for all other national defense information. *Top secret information* is defined as information that, if disclosed, "reasonably could be expected to cause exceptionally grave damage to the national security" (from Executive Order 12356).

In an offense involving false identification documents or access devices, an upward departure may be warranted where the actual loss does not adequately reflect the seriousness of the conduct.

An offense shall be deemed to have "substantially jeopardized the safety and soundness of a financial institution" (subsection (b)(6)(a)) if, as a consequence of the offense, the institution became insolvent; substantially reduced benefits to pensioners or insureds; was unable on demand to refund fully any deposit, payment, or investment; was so depleted of its assets as to be forced to merge with another institution in order to continue active operations; or was placed in substantial jeopardy of any of these.

Gross receipts from the offense (subsection (b)(6)(b)) includes all property, real or personal, tangible or intangible, obtained directly or indirectly as a result of such offense.

Criteria for Upward or Downward Adjustments and Departures

Adjustments to the offense level are based on the role the defendant played in committing the offense. The determination of a defendant's role is to be made on the basis of all conduct within the scope of relevant conduct section of the guidelines. Several categories applicable to computer crime follow:

> *Abuse of trust or use of special skill*—If the defendant abused a position of public or private trust or used a special skill in a manner that significantly facilitated the commission or concealment of the offense, the offense increases by two levels. *Public or private trust* refers to a position of trust characterized by professional or managerial discretion. *Special skill* refers to a skill not possessed by members of the general public and usually requires substantial education, training, or licensing. In ® *U.S. v. Lavin,* the federal appellate court in New York City upheld a district court's imposition of a special skills sentencing enhancement where the defendant

had installed electronic equipment in ATMs to obtain PINs and account numbers of bank customers. The court said the defendant used skills that were not possessed by the general public and that these skills greatly facilitated his crime.

Obstruction of justice—If a defendant willfully obstructs or attempts to obstruct the administration of justice, at any stage, the offense level is increased by two levels. Obstruction of justice can include a range of actions, from making false statements, and threatening witnesses to destruction of evidence.

Groups of closely related counts—All counts involving substantially the same harm are grouped together into a single group. This would cover counts involving the same victim, act, transaction, or conduct. Computer-related fraud comes under this subsection.

Criminal history—A record of a defendant's past criminal conduct, such as number and type of convictions, recidivism, and patterns of career criminal behavior are evaluated and given points to determine the criminal history category in the sentencing table.

Acceptance of responsibility—If the defendant clearly demonstrates acceptance of responsibility for his or her offense, the offense decreases by two levels.

Substantial assistance to authorities—Upon motion of the government stating that the defendant has provided substantial assistance in the investigation or prosecution of another person who has committed an offense, the court may depart from the guidelines. *Substantial weight should be given to the government's evaluation of the extent of the defendant's assistance* means essentially that the prosecutor has a big weapon in determining the fate of the defendant.

Determining the Sentence

The defendant's sentence is based on the combined offense level, which is subject to adjustments from the categories just discussed, and from the total points that determine the criminal history category in the sentencing table.

A defendant's record is relevant to sentencing because "a defendant with a record of prior criminal behavior is more culpable than a first offender and thus deserving of greater punishment." Section 4A1.1 of the guidelines gives items and corresponding points that determine the criminal history category in the sentencing table. Points are tallied for each prior sentence of imprisonment, whether the defendant committed a crime while under any criminal justice sentence, or committed a crime within two years after release from prison, or was convicted for a crime of violence.

Section 4A1.3 covers departures based on how well the criminal history category reflects the seriousness of the defendant's past criminal conduct or the likelihood that the defendant will commit other crimes.

The Sentencing Table

In Table 14-1, the Offense Level forms the vertical axis; the Criminal History Category (I-VI), the horizontal axis. The intersection of the Offense Level and Criminal History Category displays the guideline range in months of imprisonment.

Sentencing Guidelines for Organizations

The federal sentencing guidelines have the potential of affecting virtually every aspect of an organization, its operations, and its personnel. The guidelines impose greatly increased fines and penalties, "so that the sanctions imposed upon organizations and their agents, taken together, will provide just punishment, adequate deterrence, and incentives for organizations to maintain internal mechanisms for preventing, detecting, and reporting criminal conduct." This mandate for compliance programs that call for voluntary disclosure of wrongdoing, internal controls, compliance audits, and investigations is covered next.

The federal sentencing guidelines for organizations became effective November 1, 1991. The sentencing guidelines have been prepared by the United States Sentencing Commission established by the Sentencing Reform Act of 1984. The commission was charged with drafting guidelines for federal judges to use when sentencing convicted defendants. The basic premise for organizational sentencing, according to the commission, is that "organizations can act only through agents and, under federal criminal law, generally are vicariously liable for offenses committed by their agents. Federal prosecutions of organizations therefore frequently involve individual and organizational co-defendants."

The commission also has the authority to issue general policy statements concerning imposition of fines, establish the permissible width of a guideline range calling for a term of imprisonment, and provide appellate review of sentences.

The guidelines give courts direction on how to structure fines and sentences for corporations and their personnel. Organizations convicted of certain crimes can face fines significantly greater than those imposed in the past. The commission has stated that it intends to allow "the most serious criminal conduct committed by organizations to be punished at or near the statutory maximum levels established by Congress . . . and accommodate the highest fines historically imposed on organizations." Further, "the court should require the organization to take all appropriate steps to provide compensation to victims and otherwise remedy the harm caused or threatened by the offense."

The guidelines do offer flexibility for judges to apply reasoned judgment to often complex cases. And there are ways an organization can reduce the costs of fines.

This chapter examines the penalty provisions and definitions in the sentencing guidelines related to organizations and officers. The sentencing commission stated four general principles guiding organizational sentencing:

1. The court must, whenever practicable, order the organization to remedy any harm caused by the offense; that is, "as a means for making victims whole for the harm caused."

2. If the organization operated primarily for a criminal purpose, the fine should be high enough to divest the organization of all its assets.

3. The fine range should be based on the seriousness of the offense and the culpability of the organization.

4. Probation should be ordered for an organization if needed to ensure that another sanction will be fully implemented or to ensure that steps will be taken within the organization to reduce the likelihood of future criminal conduct.

Key Definitions

High-level personnel of the organization means individuals who have substantial control over the organization or who play a substantial role in setting policy within the organization. The term includes directors, executive officers, individuals in charge of a major business or functional unit of the organization, and individuals with a substantial ownership interest.

Substantial authority personnel means individuals who, within the scope of their authority, exercise a substantial measure of discretion in acting on behalf of an organization. The term includes high-level personnel, individuals who exercise substantial supervisory authority, and any other individuals who, although not part of an organization's management, nevertheless exercise substantial discretion when acting within the scope of their authority.

Agent means any individual authorized to act on behalf of the organization.

Unit of the organization means any reasonably distinct operational component of the organization.

Offense means the offense of conviction and all relevant conduct.

An individual *condoned an offense* if the individual knew of the offense and did not take reasonable steps to prevent or terminate it.

Pecuniary gain means the additional before-tax profit to the defendant resulting from the relevant conduct of the offense. Gain can result from either additional revenue or cost savings.

Net assets means the assets remaining after payment of all legitimate claims against assets by known innocent bona fide creditors.

Similar misconduct means prior conduct that is similar in nature to the conduct underlying the instant offense, without regard to whether or not such conduct violated the same statutory provision.

An individual was *willfully ignorant of the offense* if the individual did not investigate the possible occurrence of unlawful conduct despite knowledge of circumstances that would lead a reasonable person to investigate whether unlawful conduct had occurred.

Due diligence is "such a measure of prudence, activity, or assiduity, as is properly to be expected from, and ordinarily exercised by, a reasonable and prudent man under the particular circumstances; not measured by any absolute standard, but depending on the relative facts of the special case" (*Black's Law Dictionary*).

Determining the Fine

If the court determines that the organization operated primarily for a criminal purpose or by criminal means, the fine should be high enough to divest the organization of all its net assets—those assets left after all legitimate claims have been paid.

Base Fine

The base fine is the greatest of the amount from the offense level fine table corresponding to the offense level; or the pecuniary gain to the organization from the offense; or the pecuniary loss from the offense caused by the organization to the extent the loss was caused intentionally, knowingly, or recklessly. The fine applies to each count for the applicable offense. In certain cases, special instructions for determining the loss, gain, or offense level amount apply.

In most instances, the base fine measures the seriousness of the offense. Base fine determinates are selected so that, "in conjunction with the multipliers derived from the culpability score, they will result in guideline fine ranges appropriate to deter organizational criminal conduct and to provide incentives for organizations to maintain internal mechanisms for deterring, detecting, and reporting criminal conduct."

There are 32 offense levels that measure the gravity of the offense and the organization's culpability. These offenses, however, are by no means all of the crimes of which an organization might be convicted.

Offense Level Fine Table

Offense Level	Amount
6 or less	$5,000
7	$7,500
8	$10,000
9	$15,000
10	$20,000
11	$30,000
12	$40,000
13	$60,000
14	$85,000
15	$125,000
16	$175,000
17	$250,000
18	$350,000
19	$500,000
20	$650,000
21	$910,000
22	$1.2 million
23	$1.6 million
24	$2.1 million
25	$2.8 million
26	$3.7 million
27	$4.8 million
28	$6.3 million
29	$8.1 million
30	$10.5 million

31	$13.5 million
32	$17.5 million
33	$22 million
34	$28.5 million
35	$36 million
36	$45.5 million
37	$57.5 million
38 or more	$72.5 million

Offense Count Guideline

Offense Levels and Offense Descriptions

The offenses that follow are described in Chapter 2 of the sentencing commission guidelines. The letters and numbers in parentheses after each offense refer the specific section of Chapter 2.

Level 6 or less consists of the following offenses:

- Larceny, embezzlement, or other forms of theft (2B1.1);
- Receiving, transporting, transmitting, or possessing stolen property (2B1.2);
- Property damage or destruction (2B1.3);
- Trespass (2B2.3);
- Criminal infringement of a copyright (2B5.3);
- Criminal infringement of a trademark (2B5.4);
- Payment or receipt of unauthorized compensation (2C1.6);
- Illegal use of a registered number to unlawfully manufacture, distribute, acquire, or dispense a controlled substance (2D3.1);
- Manufacture of a controlled substance in excess of or unauthorized by registration quota (2D3.2);
- Illegal transportation or transshipment of a controlled substance (2D3.4);
- Miscellaneous gambling offenses (2E3.3);
- Theft or embezzlement from employee pension and welfare benefit plans (2E5.2);
- False statements and concealment of facts in relation to documents requirements under the ERISA (2E5.3);
- Embezzlement or theft from private sector labor unions (2E5.4);
- Failure to maintain or falsifying of records required by the Labor Management Reporting and Disclosure Act (2E5.5);
- Fraud and deceit (2F1.1);

- Importing, mailing, or transporting obscene matter (2G3.1);
- Failure to report theft of explosive materials (2K1.1);
- Improper storage of explosive materials (2K1.2);
- Odometer laws and regulations (2N3.1);
- Structuring transactions to evade reporting requirements (2S1.3);
- Fraudulent tax returns, statements, or other documents (2T1.5);
- Failure to deposit collected tax in a trust account as required after notification (2T1.7);
- Offenses related to withholding statements (2T1.8);
- Tax regulatory offenses (2T2.2).

Level 7 includes offering, giving, soliciting, or receiving a gratuity (2C1.2) and offering, giving, soliciting, or receiving a loan or gratuity to a bank examiner or gratuity for procuring a bank loan (2C1.6).

Level 8 includes

- Bribery in the procurement of a bank loan and other commercial bribery (2B4.1);
- Altering or removing motor vehicle identification numbers, or trafficking in motor vehicle identifications (2B6.1);
- Insider trading (2F1.2).

Level 9 includes the following:

- Offenses involving counterfeit bearer obligations of the United States (2B5.1);
- Unlawful conduct related to contraband cigarettes (2E4.1);
- Interception of communications or eavesdropping (2H3.1);
- Failure to file report of currency and monetary instruments transactions (2S1.3).

Level 10 includes the following:

- Offering, giving, soliciting, or receiving a bribe, or extortion under color of official right (2C1.1);
- Intangible right to honest services of public officials, or conspiracy to defraud by interference with governmental functions (2C1.7);
- Offering, accepting, or soliciting a bribe or gratuity affecting the operation of an employee welfare or pension benefit plan (2E5.1);
- Prohibited payments or money lending by employees or representatives or agents to employees of labor organizations (2E5.6);

- Bid rigging, price fixing, or market allocation agreements among competitors (2R1.1).

Level 12 includes the following:

- Unlawful sale or transportation of drug paraphernalia (2D1.7);
- Illegal gambling business (2E3.1);
- Transmission of wagering information (2E3.2);
- Obstruction of justice (2J1.2).

Level 17 consists of engaging in monetary transactions in property derived from specified unlawful activity (2S1.2).

Level 19 includes RICO-related offenses (2E1.1) and misprision of a felony (range 4-19; 2X4.1).

Base level offenses include the following:

- Unlawful receipt, possession, or transporting of firearms or ammunitions or prohibited transactions involving firearms or ammunitions (2K2.1);
- Failure to report money transactions or structuring monetary transactions to evade reporting requirements (2S1.3);
- Money laundering (2S1.1);
- Tax evasion (based on tax loss) (2T1.1);
- Willful failure to file a tax return, supply information, or pay tax (2T1.2);
- Fraudulent or false statement under penalty of perjury (2T1.3);
- Aiding, assisting, procuring, counseling, or advising a tax fraud (2T1.4);
- Failing to collect or truthfully account for tax (2T1.6);
- Conspiracy to impair, impede, or defeat tax (2T1.9);
- Nonpayment of taxes (2T2.1);
- Smuggling (2T3.1);
- Receiving or trafficking in smuggled property (2T3.2);
- Conspiracy (2X1.1);
- Accessory after the fact (2X3.1).

Culpability Score

The culpability score is determined by whether the organization had an involvement in or tolerance of criminal activity, a prior history of misconduct, a violation of a judicial order or injunction or a condition of probation, or an obstruction of justice charge. The score starts with five points and additions, using the greatest amount of points, for such things as the size of the organization and the unit within which the offense was committed, along with the degree of involvement of various personnel in the offense or their tolerance of the offense.

The prior history section considers the time frame of misconduct; more points are added for several charges or acts within a five year period. Points can be subtracted from the subtotal of the culpability score if the organization has an effective program to prevent and detect violations of law or reports an offense it commits, within certain parameters, to the proper authorities.

Minimum and Maximum Multipliers

Using the culpability score, along with any applicable fine instructions, minimum and maximum fine multipliers can be determined from the table that follows. Applying the multiplier pairs to the base fine yields the lower and upper points of the fine range.

Culpability Score

	Minimum	Maximum
10 or more	2.00	4.00
9	1.80	3.60
8	1.60	3.20
7	1.40	2.80
6	1.20	2.40
5	1.00	2.00
4	0.80	1.60
3	0.60	1.20
2	0.40	0.80
1	0.20	0.40
0 or less	0.05	0.20

Guideline Fine Ranges for Organizations

The fine range is determined by multiplying the base fine by the applicable minimum or maximum multiplier.

In determining the fine range within the guideline range, the sentencing commission said the court should consider the following:

- The need for the sentence to reflect the seriousness of the offense, promote respect for the law, provide just punishment, afford adequate deterrence, and protect the public from further crimes of the organization;

- The organization's role in the offense;
- Any collateral consequences of conviction, including civil obligations arising from the organization's conduct;
- Any nonpecuniary loss caused or threatened by the offense;
- Whether the offense involved a vulnerable victim;
- Any prior record of an individual within high-level personnel of the organization who participated in, condoned, or was willfully ignorant of the criminal conduct;
- Any prior civil or criminal misconduct by the organization under Section 8C2.5(c);
- Any culpability score higher than 10 or lower than 0;
- Partial but incomplete satisfaction of the conditions for one or more of the mitigating factors set forth in the culpability score; and
- Any factor listed in 18 U.S.C., Section 3572(a).

The court also may "consider the relative importance of any factor used to determine the range, including the pecuniary loss caused by the offense, the pecuniary gain from the offense, any specific offense characteristic used to determine the offense level, and any aggravating or mitigating factor used to determine the culpability score."

Fine Range Guideline Departures

The guidelines may not have covered adequately every factor that should be considered in sentencing. The commission listed factors that may be grounds for departure, including

- An organization that has given substantial assistance to authorities in investigation or prosecution;
- An offense where there is a foreseeable risk of death or bodily injury or if the offense resulted in death or risk of bodily injury;
- If the offense is a threat to national security;
- If the offense is a threat to the environment;
- If the offense is a risk to the integrity or continued existence of a market;
- If the organization, in connection with the offense, was involved in public corruption;
- If the organization is a public entity;

- If the members or beneficiaries of the organization are the direct victims of the offense;
- If the remedial costs to the organization greatly exceed the gain received; and,
- If there is exceptional organizational culpability.

Probation Sentences for Organizations

Under the guidelines, a court can impose probation on an organization if

- It is necessary to secure restitution, enforce a remedial order, or ensure completion of community service;
- It is necessary to make sure the organization has the ability to pay a monetary penalty;
- An organization with more than 50 employees does not have an effective compliance program;
- The organization engaged in similar misconduct within five years prior to sentencing;
- Management-level personnel participated in the misconduct underlying the instant offense within five years prior to sentencing;
- The sentence is necessary to ensure that changes are made within the organization to reduce the likelihood of future criminal conduct;
- The imposed sentence does not include a fine; or
- It is necessary to accomplish a purpose of sentencing under U.S.C. 3553(a)(2).

Terms and Conditions of Probation

Organizations under a felony conviction can get a one- to five-year probation sentence; in any other case, the term will not be more than five years.

Several conditions of probation include not committing another crime during the term of probation; at least either a fine, restitution, or community service; or other conditions deemed reasonable and necessary by the court.

To ensure payment of a fine or restitution, the court can order the organization to submit periodical financial statements. In addition, the organization will have to have its books examined by outside auditors—and pay for this examination. And any material change affecting the organization must be reported to the court.

The sentencing commission's recommended conditions of probation seek to force a change in organizational behavior. The court can order the organization to pay for publicizing the nature of the offense it committed, the punishment, and what the organization intends to do to prevent the recurrence of similar offenses. Under certain statutes, the court may call for the organization to develop and submit to the court "a program to prevent and detect violations of law."

The court also will prescribe the form of notice to be given to employees and shareholders as to the nature of the compliance program. The organization, further, will have to make periodic reports to the court on the progress it is making implementing the compliance program. To monitor whether it indeed is following through on all aspects of the compliance program, the organization will have to submit to a reasonable number of regular or unannounced examinations of its books and records and interrogation of knowledgeable individuals in the organization. These examinations and interrogations will be done either by the probation officer or by experts engaged by the courts. Again, compensations and costs are to be paid by the organization.

Probation Violations

Violation of a condition of probation could lead to an extension of the term, more restrictive conditions, or a revocation of probation and resentencing of the organization. Repeated serious violations could lead to having a trustee appointed to ensure the organization complies with court orders.

Assessments and Forfeitures

Special assessments may be required by statute, along with forfeitures. The court also may order the organization to pay the costs of prosecution.

15

Shoehorn Laws
Federal Statutes Used to Prosecute Computer-Related Crime

A variety of federal laws have been or could be used to prosecute computer-related crimes. Several of these laws are very broad and have been used for purposes far beyond their original intent; and they are potent legal weapons wielded by the prosecution. The major statutes in the prosecutor's arsenal include conspiracy, mail fraud, wire fraud, the Racketeer Influenced and Corrupt Organizations Act (RICO), and the National Stolen Property Act.

Conspiracy Laws

Conspiracy law is relevant to a discussion of computer-related crime, investigation, and litigation because conspiracy violations often are the first thing prosecutors look for, as generally it is easier to obtain convictions under these statutes.

The conspiracy statutes have been used in a wide range of cases, limited, it seems, only by the imaginations of prosecutors and litigators. In dealing with a possible computer crime, however, it is wise to be practical and precise, rather than imaginative.

The following discussion will cover the general conspiracy statutes, the Racketeer Influenced and Corrupt Organizations Act, and the statutes of mail and wire fraud, the two statutes commonly merged with or underlying conspiracy. And we will show how these laws can be applied to computer-related criminal and civil charges and used to facilitate recovery for damages.

Conspiracy

The general federal conspiracy statute (18 U.S.C. 371) says, "if two or more persons conspire either to commit any offense against the United States, or to defraud the United States, or any agency thereof in any manner or for any purpose, and one or more of such persons do any act to effect the object of the conspiracy, each shall be fined not more than $10,000 or imprisoned not more than five years or both."

Conspiracy is a group crime, a group agreement, a deliberate plotting to subvert the law. The elements of the crime are (1) the knowing and willful agreement to commit a crime (2) between two or more persons and (3) an action to carry out the conspiracy.

Note that the crime of conspiracy is the agreement: this is the critical element. Parties to such illegitimate agreements need not strike a bargain in the traditional business sense—create a written contract, agreement, memo, have a meeting, or the like—to be prosecuted under the statute. It need only be shown that a person was a party to a conspirational agreement to do something unlawful, during the existence of the conspiracy and that he or she knew what was involved and was committed to taking part in it to assure its success.

A person involved in a conspiracy is responsible for all that happened during the conspiracy to which he or she agreed; that is, the acts of each of the confederates in the conspiracy even though he or she may be unaware of their actions.

A member of a conspiracy may get out or withdraw only by doing something to disavow or defeat the purpose of the conspiracy. However, the person will remain liable for anything that happened prior to that withdrawal.

Evidence

It must be demonstrated that a person had knowledge of the conspiracy, agreed to it, and performed some act to further the conspiracy. Statements and acts of coconspirators are admissible as evidence if they indicate the preceding.

Prosecution

To prosecute conspiracy, the offense must be within the court's venue. Usually, conspiracy is merged with another crime, because it normally is discovered after a crime has been committed. Therefore, the charges are conspiracy plus a crime. In common law and some state statutes, the charges are merged. The prosecution also must determine if there was a single conspiracy or if there were multiple conspiracies. A single conspiracy is a single agreement, although it may involve many persons, be complex, and continue over a long period of time.

Prosecutors are aware of the advantages of charging the conspiracy offense, which includes the ease of joining charges, the admissibility of coconspirators' statements, and that it is not necessary to show that each defendant actually committed the offense—only that he or she agreed to do so.

Summary

The act of conspiracy consists of the following:

- Two or more persons must agree to commit a criminal act;
- The agreement need be only inferred;
- At least one party to the conspiracy must perform some act in furtherance of the conspiracy (this act need not be criminal);
- One party may be held responsible for the acts of his or her coconspirators even though he did not commit any substantive offenses or have actual knowledge of them—a partnership in crime includes all members within its scope and during its time of operation;
- The agreement must be to either violate a criminal statute or, under federal law, to defraud the U.S. government. The definition of fraud is not limited that of common law; the federal statute includes every conspiracy to impair, obstruct, or defeat any lawful function of the government.

Two Criminal Statutes Often Merged with Conspiracy

Mail Fraud (18 U.S.C. 1341)

The Mail Fraud statute provides that

Whoever, having devised or intending to devise any scheme or artifice to defraud, or for obtaining money or property by means of false or fraudulent pretenses, representations, or promises, or to sell, dispose of, loan, exchange, alter, give away, distribute, supply, or furnish or procure for unlawful use any counterfeit or spurious coin, obligation, security, or other article, or anything represented to be or intimated or held out to be such counterfeit or spurious article, for the purpose of executing such scheme or artifice or attempting to do so, places in any post office or authorized depository for mail matter, any matter or thing whatever to be sent or delivered by the Postal Service or takes or receives therefrom, any such matter or thing, or knowingly causes to be delivered by mail according to the direction thereon, or at the place at which it is directed to be delivered by the person to whom it is addressed, any such matter or thing, shall be fined not more than $1,000 or imprisoned not more than five years, or both.

Wire Fraud (18 U.S.C. 1343)

The essential elements of wire fraud are

1. The devising of a scheme and artifice to defraud.

2. A transmittal in interstate or foreign commerce by means of wire, radio, or television communication of writings, signs, signals, pictures, or sounds for the purpose of executing the scheme and artifice to defraud. Here is where transmission of computer data over telephone lines fits.

The statute provides that "Whoever, having devised or intending to devise any scheme or artifice to defraud, or for obtaining money or property by means of false or fraudulent pretenses, representations or promises, transmits or causes to be transmitted by means of wire, radio, or television communication in interstate or foreign commerce, any writings, signs, signals, pictures, or sounds for the purpose of executing such scheme or artifice, shall be fined not more than $1,000 or imprisoned not more than five years, or both."

The Federal Antiracketeering Law

The federal antiracketeering law, better known as the RICO (Racketeer Influenced and Corrupt Organizations) Act, has become one of the main legal weapons against a host of crimes. However, its wide use by prosecutors has caused private sector opposition to RICO, claiming the law is being used indiscriminately, harshly, and in ways not originally intended. It has been argued that RICO was to apply to criminals and racketeers—as its title says—not for civil suits against corporations with no criminal record. Specifically, the law's triple damages award has been seen as too severe for most civil cases.

Prosecution Targets

The RICO statutes (18 U.S.C., Sect. 1961-1968) originally were intended to give the government a legal tool to use in prosecuting organized crime. Since it was passed in 1970, however, the RICO Act has been used to prosecute Mafia figures, members of the Hells Angels motorcycle club, a former governor, commodity futures traders, securities brokers and firms, and antiabortion protesters.

Key Provisions of the RICO Statutes

The RICO Act does have important application to fraud and computer crime cases, and to understand the statutes, one must be cognizant of several key phrases and terms:

> *Enterprise*—Section 1961(4) defines this as any individual, partnership, corporation, association, or other legal entity and any union or group of individuals associated in fact although not a legal entity.
>
> *Pattern of racketeering activity*—This term encompasses both the act and the pattern of racketeering. The act of racketeering is defined by those statutes, state and federal, that are broken mostly in organized crime activity; the statute enumerates 32 such offenses. Included under state law are arson, bribery, extortion, dealing in narcotics or dangerous drugs, gambling, kidnapping, murder, and robbery. Under Title 18, U.S.C. are Sections 201 (bribery), 224 (sports bribery), 471-473 (counterfeiting), 659 (theft from interstate shipment), 664 (embezzlement from pension and welfare funds), 891-894 (extortionate credit transactions), 1084 (transmission of gambling information), 1341 (mail fraud), 1343 (wire fraud), 1503 (obstruction of justice), 1510 (obstruction of criminal investigations), 1511 (obstruction of state or local law enforcement), 1951 (interference with commerce, robbery or extortion), 1952 (racketeering), 1953 (transportation of wagering paraphernalia), 1954

(unlawful welfare fund payments), 1955 (prohibition of illegal gambling businesses), 2314-2315 (interstate transportation of stolen property), and 2421-2424 (white slave traffic).

Under Title 29, U.S.C. are Sections 186 (restrictions on payments and loans to labor organizations) and 501(c) (embezzlement from union funds). Also included are offenses punishable under any U.S. law involving bankruptcy fraud, felonious manufacture or other dealing in narcotic or dangerous drugs, or securities fraud.

Under the Financial Institutions Reform, Recovery and Enforcement Act of 1989 (PL 101-73), the RICO Act is applied to bank fraud. Environmental crimes and crimes against federal interest computers now fall under the RICO Act.

Convictions Under the RICO Act

For conviction under the RICO Act, at least two of the preceding acts of racketeering (to form a pattern) must have a nexus, or connection, with the enterprise itself. Therefore, for the government to convict under the RICO Act, it must prove both the existence of an enterprise and the connected pattern of racketeering activity. Of the two acts of racketeering, the most recent must have occurred within 5 years of the indictment and another within 10 years of the most recent act.

Penalties Under RICO

The RICO Act provides both criminal penalties and civil remedies. Criminal penalties are 20 years imprisonment, a $25,000 fine, and forfeiture to the government of any interest in, or property in contractual right to, any enterprise acquired in violation of Section 1962.

Civil remedies allowable under the RICO Act are similar to those in antitrust law; in addition, any person injured by the enterprise may sue and recover three times the damage sustained plus court costs and reasonable attorneys' fees.

Cases and Court Decisions

In its criminal context, the RICO Act has been upheld by the Supreme Court (see *Russello v. United States*) and cases that have been successfully prosecuted include charges of arson, extortion, mail fraud, racketeering, and conspiracies to commit such crimes. Also, the U.S. Supreme Court let stand a ruling by the Seventh Circuit Court of Appeals (in

Schact v. *Brown*) that indicates upholding the civil side of the RICO Act. The Court said such suits could go forward even if the defendant had not previously been convicted of a "mobster-type crime."

RICO Decision May Offer a New Legal Weapon Against Violent Protesters

The Supreme Court's 1994 ruling in a case involving antiabortion protesters charged under provisions of the RICO statute could be used in future prosecutions against groups that threaten or use violence against individuals or organizations. Protest groups that engage in lawful, nonviolent protest should be protected under the First Amendment, but groups that use the threat of or actual violence to further their cause may now be charged as an "enterprise" that engages in a "pattern of racketeering" under the RICO Act.

We next examine this case under the RICO Act as well as the extortion statute and its application to prosecution of extremist groups employing violence and extortion to further their political, religious or social goals. Of course, many protesters have used tactics that come under malicious mischief or breaking and entering laws. Other protesters, however, have wreaked destruction on laboratories and computer centers, and threatened and intimidated professionals into abandoning research programs.

In *National Organization for Women, Inc., etc., et al., v. Joseph Scheidler et al.* (No. 92-780), the Supreme Court accepted a writ of certiorari brought by petitioners, the National Organization for Women, Inc., and two health care centers that perform abortions. In the suit, the respondents were a coalition of antiabortion groups called the Pro-Life Action Network (PLAN) and others. The petitioners sued alleging violations of the Racketeer Influenced and Corrupt Organizations statute; an amended complaint added violations of the Hobbs Act.

The petitioners claimed that the respondents were members of a nationwide conspiracy to shut down abortion clinics through a pattern of racketeering activity including extortion. The proabortion groups sought injunctive relief, along with treble damages, costs, and attorneys' fees.

Both the federal district court and the court of appeals had dismissed the case. These courts said that the voluntary contributions received by the antiabortion groups did not constitute income derived from racketeering activities for purposes of the RICO Act. Further, that "non-economic crimes committed in furtherance of non-economic motives are not within the ambit of RICO."

The Supreme Court took the case to resolve a conflict among the courts on the putative economic motive requirement of specific RICO provisions, 1962(c) and (d); that is, whether the racketeering enterprise or the racketeering predicate acts must be accompanied by an underlying economic motive.

Nowhere in the act's definitions could the Supreme Court find "any indication that an economic motive is required." Subsection 1962(c) may suggest that there be a profit-seeking motive, but the Court argued that *enterprise* in this subsection "connotes generally the vehicle through which the unlawful pattern of racketeering activity is committed, rather than the victim of that activity . . . Consequently, since the enterprise in subsection (c) is not being acquired, it need not have a property interest that can be acquired nor an economic motive for engaging in illegal activity; it need only be an association in fact that engages in a pattern of racketeering activity. Nothing in subsections (a) and (b) directs us to a contrary conclusion."

Nor do predicate acts require an underlying economic motive. "Predicate acts, such as the alleged extortion here, may not benefit the protesters financially, but they may still drain money from the economy by harming businesses such as the clinics." The Court thought the requirement of an economic motive "neither expressed nor . . . fairly implied in the operative sections of the Act."

Summary

In this case on specific provisions in the RICO statute, the Supreme Court ruled no proof is needed that either (1) the "racketeering enterprise" definition or (2) the underlying predicate acts of racketeering were motivated by an economic purpose. *Enterprise* means an "association in fact" regardless of motive.

The Court's ruling applies to both civil and criminal cases.

The Court, while showing great concern for First Amendment rights and challenges to the RICO Act in future cases, indicated that some acts of so-called ideological expression needed chilling when too violent.

The Hobbs Act

In *NOW* v. *Scheidler*, the Supreme Court did not reach the issue of the antiabortion coalition violating the Hobbs Act; the predicate acts in the case were conspiracy and extortion. NOW and the other petitioners claimed the antiabortion coalition constituted "members of a nationwide conspiracy to shut down abortion clinics . . . and that they conspired to use threatened or actual force, violence or fear to induce clinic employees, doctors, and patients to give up their jobs, their right to practice medicine, and their right to obtain clinic services; that the conspiracy injured the clinics' business and property interests."

Although the Court did not rule on the issue of extortion, it is worth examining how a charge of extortion might be invoked and under what circumstances. First, we should look at how the law reads.

The Hobbs Act (18 U.S.C., Sect. 1951(a)) provides that "Whoever in any way or degree obstructs, delays, or affects commerce or the movement of any article or commodity in commerce, by robbery or extortion or attempts or conspires so to do, or commits or threatens physical violence to any person or property in furtherance of a plan or purpose to do anything in violation of this section shall be fined not more than $10,000 or imprisoned not more than 20 years or both."

Section 1951(b)(2) defines *extortion* as "the obtaining of property from another, with his consent, induced by wrongful use of actual or threatened force, violence, or fear, or under color of official right."

The Hobbs Act has two essential elements: obstruction or interference with interstate commerce and extortion. Interference with interstate commerce may be actual or potential. Extortion may be threatened or actual, but there must be acts, the natural effect of which instills fear in the victim. To establish a violation of the Hobbs Act there must be a loss to the victim, property or "any valuable right considered as a source or element of wealth."

The definition of property includes intangible as well as tangible property. Intellectual property, computer data, and information could fall under property; in accounting terms, these are intangible assets, which also can include organization costs, leaseholds, and goodwill.

The Travel Act

Related to the Hobbs Act, the Travel Act (18 U.S.C., Sect. 1952) adds the element of interstate travel or use of "any facility in interstate or foreign commerce" with intent to "1) distribute the proceeds of any unlawful activity; or, 2) commit any crime of violence to further any unlawful activity; or, 3) otherwise promote, manage, establish, carry on, or facilitate the promotion, management, establishment, or carrying on, of any unlawful activity, and thereafter performs or attempts to perform any of the acts specified in subparagraphs 1, 2, and 3 shall be fined not more than $10,000 or imprisoned for not more than five years, or both."

A Hobbs Act violation has been described as an essentially local criminal act, transformed by its impact, into an interstate and federal offense. A Travel Act violation basically is a local offense plus the use of an interstate facility. Both acts were originally designed to control organized crime that crossed state boundaries in furtherance of criminal activity.

RICO and Extremist Groups

For computer users who have been the targets of protesters or extremist organizations, *NOW* v. *Scheidler* may offer new legal responses to threats or violent actions.

It is a fact that some protesters have learned from and use strategies and tactics first practiced and codified by terrorists. By this, I mean acts that are far removed from lawful, nonviolent protest; and I am referring to an extremist group that threatens or uses violence against individuals, groups, and organizations as a major coercive tool. Extremists use terrorism, which can be defined as the systematic use or threat of violence, to achieve political or social goals. This use of violence as coercion is supported by propaganda, which is used to explain and "justify" the actions of terrorists. Terrorism is never an isolated act and is never done as an end in itself. It is a tactic of extremism and is used as one method of achieving extremists' goals.

The threat or use of violence is directed at particular targets, selected because they most suitably represent the extremist's political or social protest goals. The target is the "enemy" that must be coerced or intimidated into doing what the extremist desires. Actions against a target enemy must be evaluated for their propaganda value; a wider audience must be convinced that the extremist goal is valid. Extremists have learned to use the media to communicate their extortionate demands and to carry their propaganda to a wide audience.

Remember, terrorism is never an end in itself. It is a tool to accomplish specific goals. No act of terrorism is an isolated event; it is one manifestation of a program of extremist propaganda and coercion.

These coercive threats and acts of violence often are defined under laws on extortion, arson, murder, use of explosives, or terrorism.

Conspiracy and Wire Fraud Charges No Good Against Software Piracy

The United States District Court in Massachusetts threw out an indictment against David LaMacchia, charged with one count of conspiracy and wire fraud in a scheme to upload copyrighted software then download it free to subscribers of a bulletin board. The court said LaMacchia could not be prosecuted for criminal copyright infringement under the wire fraud law.

LaMacchia was indicted using the conspiracy statute and the commonly used wire fraud statute. The essential elements of wire fraud are the devising of a scheme and artifice to defraud and the transmittal in interstate commerce by means of wire, or the like, of writings, and such, for the purpose of executing the scheme or artifice to defraud or to obtain money or property by false pretenses.

In a criminal prosecution for copyright infringement, the government must prove that a copyright was infringed, that it was a willful violation, and that the infringement was for profit.

LaMacchia's scheme cost copyright holders more than $41 million, but LaMacchia did not seek nor receive any personal reward.

The Justice Department, it seems, was trying to test the use of the wire fraud statute in copyright cases. But the court asked "whether the bundle of rights conferred by copyright is unique and distinguishable from the indisputably broad range of property interests protected by the wire fraud statute." The court ruled that indeed copyright was unique and the wire fraud statute could not be applied (*United States* v. *David LaMacchia*).

Overview of the Civil Provisions of the RICO Act

The civil provisions of the RICO Act (18 U.S.C., Sect. 1964-1968) are predicated on the general provisions of the act in 18 U.S.C., Sections 1961 and 1962. A suit under the civil provisions of the RICO Act may be brought by either governments or private persons. The RICO Act's civil provisions offer expansive remedies, including treble damages, the costs of investigations and prosecution, attorneys' fees, and equitable relief.

Section 1964(a) gives federal district courts jurisdiction to grant injunctive and other equitable relief to prevent and restrain violations of Section 1962. Section 1964(a) authorizes courts to provide such relief by issuing appropriate orders, including, but not limited to

1. Ordering any person to divest himself or herself of any interest in an enterprise;

2. Imposing reasonable restrictions on future activities of investments of any person, including prohibiting the person from engaging in the same kind of endeavor as the enterprise engaged in;

3. Ordering dissolution or reorganization of any enterprise.

Section 1964(c) provides that "[a]ny person injured in his business or property by reason of a violation of Section 1962" may sue and recover treble damages, costs, and reasonable attorneys' fees.

Section 1964(d) provides that a final judgment or decree rendered in favor of the United States in any criminal RICO proceeding stops the defendant from denying the essential allegations of the criminal offense in a subsequent civil case under the RICO Act brought by the government. This provision is very useful to the government when civil cases under the RICO Act are filed following a criminal prosecution. Basically, this provision will prevent a defendant from contesting any of the factual allegations that were proven in the criminal proceeding. As a result, if the civil suit under the RICO Act is based on essentially the same allegations as the criminal RICO prosecution, the government should prevail on a motion for summary judgment against any defendants who were convicted in the criminal proceeding.

Sections 1965-1968 contain provisions involving procedural aspects of civil actions under the RICO Act—venue and service of process. Section 1966 provides for expedited treatment of civil RICO lawsuits brought by the government if the attorney general files with the court a certificate stating that the case is of public importance. Section 1967 provides that proceedings in or ancillary to civil RICO suits brought by the United States may be open or closed to the public "at the discretion of the court after consideration of the rights of affected persons." Section 1968 provides detailed procedures for the issuance of civil investigative demands by the United States prior to the institution of criminal or civil proceedings.

State RICO Laws

State RICO statutes often give prosecutors authority to sue on behalf of the state and its citizens who suffer a "business or property" injury. Most states have designed their civil RICO statutes on the federal RICO Act. Typical state statutes will prohibit

1. Investing the proceeds of a "pattern of racketeering activity" in an enterprise;
2. Taking control of an enterprise through a pattern of racketeering;
3. Infiltrating an enterprise and engaging in a pattern of racketeering through it; and,
4. Conspiring to do any of the preceding.

State statutes may substitute the term *criminal profiteering activity* for "racketeering activity." The underlying offenses (racketeering acts) may be a wide selection of state and federal crimes.

Again, the key terms to examine in state RICO statutes are *pattern* and *enterprise,* to see how precisely they are defined.

Legal remedies usually include treble damages, attorneys' fees and costs, civil forfeiture, and civil penalties.

Conclusion

The RICO Act, in its criminal and civil provisions, is a potent weapon in the prosecution's arsenal. The RICO Act is a complex law with harsh criminal penalties and expansive civil remedies. It is also a law with provisions that could be clarified or limited by the higher courts.

Conspiracy laws should be discussed with legal counsel at the outset of any fraud or computer fraud-related incident that might involve a violation of law. Issues of evidence gathering, discovery, merged or underlying offenses, or recovery of damages should all be thoroughly examined prior to an investigation or filing of legal charges.

National Stolen Property Act

The National Stolen Property Act (18 U.S.C., Sect. 2314) calls for criminal sanctions against any person who "transports, transmits, or transfers in interstate or foreign commerce any goods, wares, merchandise, securities or money, of the value of $5,000 or more, knowing the same to have been stolen, converted or taken by fraud . . . " Penalties can be a fine of up to $10,000 or a prison sentence of 10 years or both.

Federal courts have ruled that confidential information stored on a computer is valuable property under the definition of "goods, wares, or merchandise" and a person who "transmitted" stolen proprietary business information from one computer to another across state lines could be prosecuted under the statute.

The key clauses of the statute that must be satisfied for a conviction are

1. The items must be transported or transmitted in interstate or foreign commerce.
2. The items must meet the definition of goods, wares, merchandise, securities, or money.

3. The items—property or money—must have a value of $5,000 or more.
4. The defendant must have knowledge that the items were stolen or falsely made.
5. The items must have been stolen, converted, or taken by fraudulent means.

The aim of the statute is to prohibit the use of interstate transportation facilities to move stolen goods and to punish theft of property that was beyond the capability of an individual state. Therefore, the "movement," for example, of stolen confidential proprietary information, such as a trade secret, across state lines must be established (see *United States* v. *Riggs*). Also, it must be established that the defendant knew that the information was property and that it was stolen. The act also reaches fraudulent transfers of funds (see *United States* v. *Kroh*).

Obviously, other federal laws come into play when stolen information or money is transferred or transmitted. For example, the wire or mail fraud statutes often are merged with or underlie a charge of conspiracy.

The "goods and wares" clause of the act has been applied to computer software stored on disk or tape.

Possible Legal Remedy for Unsolicited Fax Ads

A legal remedy for getting unsolicited advertising via fax that can tie up a computer system may lie in the Telephone Consumer Protection Act of 1991 (TCPA) (47 U.S.C.S., Sect. 227, and collateral Code of Federal Regulations 47 C.F.R., Sect. 64.1200). The TCPA defines a *fax* as the equivalent of e-mail; that is, if your computer has a modem connected to a regular telephone line and a printer connected to that computer. An unsolicited advertising sent via e-mail to a fax is the same.

The TCPA allows a private right of action against the sender of unsolicited advertising, provided the company did not have a prior "established business relationship" with the sender. The recipient can sue for $500 for each violation, or actual monetary loss, whichever is greater, or both such actions. An injunction action also is available. If the court finds the defendant willfully or knowingly violated the TCPA, the court has the discretion of tripling the damage award.

The TCPA defines *unsolicited advertising* as "any material advertising the commercial availability or quality of any property, goods or services which are transmitted . . . "

Congress has stated that the purpose of the TCPA is to "facilitate interstate commerce by restricting certain uses of fax machines." Congress saw that certain uses of fax machines could tie up business phone lines, access time to download messages and read them, and impose a cost on the called party in having to pay for the fax paper used or even the call itself.

The TCPA was felt to be consistent with the First Amendment in that the TCPA does not restrict or discriminate messages based on their content; it regulates and restricts only the manner and place.

16

State Computer Crime Statutes

State statutes tend either to be comprehensive and deal exclusively with computer-related crime by defining specific computer elements and offensive acts or computer-related offenses to merge with existing statutes covering established crimes.

Crimes involving computers have and continue to be prosecuted using "shoehorn" laws, such as embezzlement, larceny, malicious mischief, and fraud; in the federal courts, mail or wire fraud, interstate transportation of stolen property, and conspiracy statutes commonly are used. However, many legislatures felt the need for an additional statute that proscribed various forms of computer abuse.

This chapter will review the typical elements in state computer crime statutes. The objective is to provide a method for determining whether to use a state statute in prosecuting an offense, to guide an investigation, or to select an alternative statute or a common law solution. The question format can be helpful in reviewing your state law on computer crime and perhaps fitting it to a legal issue. Following a review of statute elements is a list of state statutes with code section locations.

Elements to Examine in a State Computer Crime Statute

How does the statute define the following terms?

- Access,
- Computer,
- Computer hacking,
- Computer network,
- Computer program,
- Computer software,
- Computer system,
- Criminal intent (willfully, knowingly, and without authorization, etc.),
- Data,
- Financial instrument,
- "Intent to permanently deprive" (a computer owner of . . .),
- Property,
- Thing of value.

How are the following offenses described?

- Access to computer system examine files (voyeurism),
- Access to defraud,
- Access to obtain money,
- Aiding and abetting (use of a computer to facilitate the commission of a crime),
- Alteration (of program, communication),
- Altering of financial instrument,
- Computer contaminant (worm or virus),
- Computer fraud,
- Computer tampering,
- Conspiracy to engage (in criminal activity),
- Damage to a computer,
- Deprived of the use of (computer or computer services),
- Destruction of a computer or information,
- Extortion,
- False pretense,
- Forgery or instrument of forgery,
- Fraudulent representation or misrepresentation,
- Fraudulent scheme,
- Modify (equipment, supplies, data),
- Offenses against computer equipment and supplies,
- Offenses against computer users,
- Offenses against intellectual property,
- Pattern of criminal activity,
- Taking possession (unauthorized taking control over a computer system through the use of "drop-dead" devices),

- Theft of services,
- Transmission of false data,
- Unauthorized access,
- Unauthorized copying (of computer files or programs),
- Unauthorized or unlawful computer use.

Describe the degree of criminal liability in the statute:

- Class of felony (A, B, C),
- Class of misdemeanor (A, B, C),
- Maximum penalty,
- Minimum penalty,
- Maximum fine,
- Minimum fine,
- Valuations and amounts of damage = levels of sanction/penalty,
- Forfeiture,
- Offender restrictions and other special provisions.

Describe the venue provisions; if different from where the crime was committed, list places where offense could be prosecuted, such as the computer owner's main place of business.

Describe any civil remedies specifically provided under the statute:

- Injunctive relief,
- Compensatory damages,
- Punitive damages,
- Attorneys' fees,
- Civil forfeiture.

Describe additional features of the statute:

- The statute of limitations in bringing a crime to a formal charge,
- Proof of injury,
- Extent of damage—all costs involved in computer and information restoration and in determining the extent of damage,
- Audits necessary to determine extent of loss and replacement,
- Parent or guardian responsible for the acts of unemancipated minor,
- Duty to report computer-related offense to law enforcement or regulatory agency,
- Specific evidence rules for computer materials.

Preemption of State Statutes

The Copyright Act of 1976 attaches federal copyright protection to a work the moment it is fixed in tangible form. The federal government also has complete responsibility for enforcing copyright law. The key provision on preemption is Section 301 of the Copyright Act of 1976, which ends copyright protections under the common law or statutes of any state. Title 28 U.S.C., Section 1338, makes it clear that any action involving rights under the federal copyright law come within the exclusive jurisdiction of the federal courts.

For a state law to be preempted, two conditions must be met:

1. The state right must be "equivalent to any of the exclusive rights within the general scope of copyright as specified by section 106."

2. The right must be "in works of authorship that are fixed in a tangible medium of expression and come within the subject matter of copyright as specified by sections 102 and 103" (see the previous condition).

Two general areas are left unaffected by the preemption:

1. Subject matter that does not come within the subject matter of copyright;
2. Violations of rights that are not equivalent to any of the exclusive rights under copyright.

States may not add to their criminal statutes an additional element that creates a distinguishable offense from the proscriptions in the copyright act.

Complete Preemption and Computer Crime Laws

A state cause of action can be preempted if the work at issue is copyright material and that state law created an equivalent right within the general scope of copyright. State computer crime laws may be in violation of the U.S. Constitution and preempted if property is defined to include copyrighted information or criminal acts that involve copyright material. Cases in state courts could be dismissed.

State Computer Crime Statutes and Code Location

Alabama	Ala. Code 13A-8-101 to 103
Alaska	Alas. Stat. Sect. 11.46.200(a) and 11.46.740
Arizona	Ariz. Rev. Stat. Sect. 13-2316
Arkansas	Ark. Stat. Sect. 5-41-101 to 107
California	Cal. Penal Code Sect. 502
Colorado	Colo. Rev. Stat. Sect. 18-5.5-101 to 102
Connecticut	Conn. Gen. Stat. Ann. Sect. 53a-250 to 261
Delaware	Del. Code Tit. 11, Sect. 931-939
Florida	Fla. Stat. Ann. Sect. 815.01 to 815.07
Georgia	Ga. Code Ann. Sect. 16-9-90 to 95
Hawaii	Haw. Rev. Stat. 708-890 to 896
Idaho	Idaho Code Sect. 18-2201 to 2202
Illinois	Ill. Stat. Ann. Ch. 38, Sect. 16D to 7
Indiana	IC 35-43-1-4 and 35-43-2-3
Iowa	Iowa Code Ann. Sect. 716A.1 to 16
Kansas	Kans. Stat. Sect. 21-3755
Kentucky	Ky Rev. Stat. Sect. 434.840 to 860
Louisiana	La. Rev. Stat. 14:73.1 to 5
Maine	Me. Rev. Stat. Ann. Tit. 17-A Sect. 431-433
Maryland	Md. Ann. Code Art. 27, Sect. 146
Massachusetts	Mass. Gen Laws Ann. Ch. 266, Sect. 30(2)
Michigan	Mich. Comp. Laws Ann. Sect. 752.791 to 797
Minnesota	Minn. Stat. Ann. Sect. 609.87 to 891
Mississippi	Miss. Code Ann. Sect. 97-45-1 to 13
Missouri	Mo. Ann. Stat. Sect. 569.093 to 099
Montana	Mont. Code Ann. 45-6-310 to 311
Nebraska	Neb. Rev. Stat. Sect. 28-1343 to 1348
Nevada	Nev. Rev. Stat. Sect. 205.473 to 490
New Hampshire	N.H. Rev. Stat. Ann. Sect. 638:16 to 19
New Jersey	N.J. Rev. Stat. Sect. 2C:20-23 to 34
New Mexico	N.M. Stat. Ann. Sect. 30-45-1 to 7
New York	N.Y. Penal Law Art. 156 to 156.50
North Carolina	N.C. Gen. Stat. 14-453 to 457
North Dakota	N.D. Cent. Code Sect. 12.1-06.1-08
Ohio	Ohio Rev. Code Ann. Sect. 2913.014
Oklahoma	Okla. Stat. Ann. tit. 21, Sect. 1951-1958
Oregon	Or. Rev. Stat. 164.377
Pennsylvania	Pa. Stat. Ann. tit. 18, Sect. 3933
Rhode Island	R.I. Gen. Laws Sect. 11-52-2 to 5
South Carolina	S.C. Code Sect. 16-16-10 to 40

South Dakota	S.D. Codified Laws Ann. Sect. 43-43B-1 to 8
Tennessee	Tenn. Code Ann. Sect. 39-14-601 to 603
Texas	Tex. Penal Code Sect. 33.01-33.05
Utah	Utah Code Ann. Sect. 76-6-701 to 705
Virginia	Va. Code Ann. Sect. 18.2-152.1 to 14
Washington	Wash. Rev. Code Ann. Sect. 9A.52.110 to 130
West Virginia	W.Va. Code Sect. 61-3c-1 to 21
Wisconsin	Wis. Stat. Ann. Sect. 943.70
Wyoming	Wyo. Stat. Sect. 6-3-501 to 505

Discovery and Computer Evidence

Information Control and Access Strategies

In litigation, the prosecution and defense have distinct strategies related to information. The prosecution strategy is to gain access to as much information as possible and usually as quickly as possible. The primary legal techniques used by a private sector plaintiff are the discovery procedures of interrogatories and motions or requests for documents, interrogatories, and subpoenas.

Government prosecutors usually use administrative summonses, grand jury subpoenas, search warrants, consent searches, civil investigative demands, and parallel proceedings.

For the defense, the strategy is to control access to critical information. The primary legal methods of controlling access are the attorney-client privilege and the work product rule.

Purposes of Discovery

The primary purpose of discovery is to obtain the information necessary to establish the allegations made in a pleading. At the outset, the party asserting an injury does not have to have all the evidence to make the case. Through discovery, the plaintiff may

- Gather the evidence necessary to support the allegations;
- Verify information already known or suspected;

- Expose to the defense its weaknesses and perhaps force a favorable settlement.

Proving a Negative

With the rapid growth of compliance litigation, records are critical to establish that you did not do something wrong. In civil litigation, the defendant, individual or organization, must prove by a preponderance of evidence that it met the requirements of the regulation.

To prove that you did not violate a regulation, you must present documented proof that will reflect a pattern of behavior in accordance with the regulation. Documented proof means information such as written policies, compliance audits, investigative reports, memos, and other items that reveal the actions of the corporation's officers, management, and personnel.

Case Theory and Evidence

Most investigations have a theory of the case at the outset. This usually means there has been a review of the applicability of particular criminal statutes to the initially known facts of the case. During information and evidence gathering, the theory of the case may be modified or develop in a different direction. This is simply a determination, based on new information, of which violations are demonstrated most clearly by the evidence gathered, what additional evidence maybe required, and what evidence might be needed to negate defenses.

Information Sources

Information comes from either persons or things. In internal corporate investigations, things usually are business records and recorded information about specific activities of personnel and executives. Information also can come from persons inside and outside the business. To gather information most efficiently and thoroughly, it is vital for the investigator to know the system of the organization—its paper flow, disposition of documents, its procedures for claims, payments, and so forth—and personnel and job functions, to know who is most likely to have what information.

Discovery Methods

Discovery encompasses the procedures available in a lawsuit for obtaining information from other litigants and third parties. These procedures are spelled out in the Federal Rules of Civil Procedure, the Federal Rules of Criminal Procedure, and by state jurisdictions.

Civil discovery methods available under the federal rules are described in Rule 26(a) as depositions upon oral examination or written questions, written interrogatories, production of documents or things, and requests for admission.

In criminal litigation, the prosecution may try to obtain business records and information through the use of subpoenas, search warrants, summonses, civil investigative demands, and consent searches, as well as depositions and interviews.

In contrast to civil discovery, requests for information are issued by an institution of the government, for example, a grand jury, which may subpoena witnesses to appear in court at a certain time, date, and place and answer questions or produce documents, records, or physical evidence.

Subpoenas

An organization responding to a subpoena to produce documents must produce them as they are kept in the usual course of business or must organize and label the records according to the demand. Subpoenas must be reasonably specific. Precise identification of exact documents is not required; a reasonable particularity will suffice.

Subpoenas can be quashed because they are too broad. The Fifth Amendment privilege against self-incrimination for individuals also may be used to quash a subpoena. A corporation, however, comes under the collective entity doctrine and is not protected by the Fifth Amendment.

Arrest Warrants

Arrest warrants may allow search and seizure of evidence and must be supported by probable cause. Warrants must be signed by a magistrate.

Search Warrants

Under the Federal Rules of Criminal Procedure, Rule 41(b), a warrant may be issued to search for and seize any "property that constitutes evidence of the commission of a criminal offense; or . . . property designed or intended for use which is or has been used as a means of committing a criminal offense."

To be issued, a search warrant must identify the property to be searched and the documents or records to be seized; that is, the warrant must satisfy the Fourth Amendment's particularity requirement.

If records are in a computer, the computer may be placed under constructive seizure until the government can understand the operating system, run the computer's programs, and generate all the documents and records specified in the search warrant.

Sample Federal Search Warrants

The following sample search warrants are taken from "Federal Guidelines for Searching and Seizing Computers," an internal report by the U.S. Department of Justice.

For E-Mail

It is essential to evaluate each case on its facts and craft the language of the warrant accordingly. Computer search warrants, even more than most others, are never one-size-fits-all products.

In some situations, you [the federal agent] may know or suspect that the target's computer is the server for an electronic bulletin board (BBS). If you need to seize the computer, the data on it, or backups of the data, consider the applicability of 18 U.S.C. Sec. 2703 (Electronic Communications Privacy Act). If the statute applies and there is or may be qualifying e-mail on the computer, consider whether the government has probable cause to believe that all or any of it is evidence of crime.

For Electronic Mail on a BBS Server

Your affiant has probable cause to believe that [the suspect]'s computer operates, in part, as the server (or communications center) of an electronic bulletin board service (BBS). This BBS [appears to] provide "electronic communication service" to other persons, and [may] contain[s] their "electronic communications," which may have been in "electronic storage" on [the suspect's] computer for less than 180 days (as those terms are defined in 18 U.S.C. Sec. 2510). The affiant is aware of the requirements of Title 18 U.S.C. Sec. 2703 describing law enforcement's obligations regarding electronic communications in temporary storage incident to transmission, as defined in that statute.

Where there may be some e-mail that could be evidence of a crime, the affidavit and warrant should distinguish and describe which will be searched and which will not. The affidavit should identify, if possible, the particular individuals by name or "hacker handle." In some cases the government may be allowed to run "string searches" of all e-mail for certain key words or phrases.

Duty to Report Laws

A number of federal and state laws and regulations require disclosure of corporate activities that are either illegal or arguably illegal. Reporting and disclosure laws often have gray areas in definitions of proscribed behavior plus requirements to disclose more than just completed formal reports. Also wanted are the "working papers" in all forms, hard copy and electronic.

Voluntary Disclosure Programs

Voluntary disclosure programs originally were developed for government contractors to encourage contractor cooperation to disclose potential or actual wrongdoing. Under a disclosure agreement, the government can demand access to records and "associated papers" and perhaps records that are more than "directly pertinent." The government may ask for an additional written agreement to facilitate the discovery of documents and evidence.

Compliance Programs

Individual agency compliance programs are similar to voluntary disclosure programs. The Federal Sentencing Commission Guidelines, however, have institutionalized the requirements for reporting and investigating illegal or possibly illegal incidents. Disclosures must be sufficient for "law enforcement personnel to identify the nature and extent of the offense and the individual(s) responsible for the criminal conduct."

Civil Investigative Demands

Under the False Claims Act and other laws, the Department of Justice is authorized to issue civil investigative demands (CIDs) for documents and testimony relevant to the law. CID authority permits access to information developed before a grand jury. However, the standards governing subpoenas and civil discovery apply to protect against disclosure of information subject to a privilege.

The Scope, Limits, and Sanctions of Discovery

Parties in litigation may obtain discovery in any matter, if it is not privileged, that is relevant to the subject matter of the pending action or that relates to the claim or the defense of one of the parties.

Using discovery is limited in frequency and extent if it is "unreasonably cumulative or duplicative" or available from another source that is more convenient or less expensive, ample opportunity already has been allowed in discovery, or the discovery is unduly burdensome or expensive. Methods used in discovery are not limited as to frequency or order of use.

A party seeking pretrial production of documents must demonstrate their relevancy to the litigation at hand, admissibility, and specificity.

Requests for documents also must be reasonable and not create undue hardship.

Documents are discoverable as long as they are not protected by either the attorney-client privilege or the work product privilege.

Destruction of discoverable evidence, intentionally, recklessly, or through negligence, can bring charges of spoliation, which is charged under obstruction of justice statutes, or court-imposed sanctions, including tort remedies; specific discovery sanctions such as contempt of court, default judgments, and impositions of costs; and instructions to juries to draw unfavorable inference against the destroyer of evidence.

Rules of Evidence

A court trial is intended to deduce the truth of a given proposition. In a criminal case, the proposition is the guilt or innocence of an accused. The evidence introduced and received by the court to prove the charge must be beyond a reasonable doubt, not necessarily to a moral certainty but that quantity and quality of evidence which would convince an honest and reasonable layperson that the defendant is guilty after all the evidence is considered and weighted impartially. The level of proof in a civil case, by contrast, requires only a "preponderance of the evidence."

But, what is evidence and how can it be weighted and introduced? In a broad sense, evidence is anything perceptible by the five senses, and any form or species of proof such as testimony of witnesses, records, documents, facts, data, or concrete objects legally presented at a trial to prove a contention and induce a belief in the minds of the court or jury. In weighting evidence, the court or jury may consider such things as the demeanor of a witness, his or her bias for or against an accused, and any relationship to the accused. So evidence can be testimonial, circumstantial, demonstrative, inferential, and even theoretical when given by a qualified expert. Evidence is simply anything that proves or disproves any matter in question.

To be *legally* acceptable as evidence, however, testimony, documents, objects, or facts must be competent, relevant, and material to the issues being litigated and be gathered in a lawful manner. Otherwise, on the motion by the opposite side, the evidence may be excluded. Now perhaps we should elaborate on relevancy, materiality, and competency:

"Relevancy of evidence does not depend upon the conclusiveness of the testimony offered, but upon its legitimate tendency to establish a controverted fact" (*ICC v. Baird*, 24 S CT. 563, 194, U.S. 25, 48 L. Ed. 860).

Some of the evidentiary matter considered relevant and therefore admissible are

1. The motive for the crime;
2. The ability of the defendant to commit the crime;
3. The opportunity to commit the crime;
4. Threats or expressions of ill will by the accused;
5. The means of committing the offense (possession of a weapon, tools, or skills used in committing the crime);
6. Physical evidence at the scene linking the accused to the crime;
7. The suspect's conduct and comments at the time of arrest;
8. The attempt to conceal identity;
9. The attempt to destroy evidence;
10. Valid confessions.

Although the motive for a crime is not an element of necessary proof to sustain a conviction (but the criminal *intent* usually is), motive is important to the investigator because it tends to identify the more likely suspects when the actual culprit is unknown. The motive also helps to construct a "theory of the case"; that is, the who, what, when, where, how, and why of the crime. So motive should not be discounted just because it is not an element of proof of a crime. Motive and motivation can narrow the search for the culprit and can be a substantial aid in reconstructing the crime (building a theory of the case.)

Supreme Court Decides Evidence from Faulty Computer Records Still Admissible

The exclusionary rule is a judge-based remedy dating from 1914, whose purpose is to deter law enforcement from violating suspects' Fourth Amendment rights by excluding evidence seized as a result of a search and seizure problem, such as a faulty warrant. In 1961, the Supreme Court, in *Mapp* v. *Ohio*, applied the rule to state courts.

In the recent case *Arizona* v. *Evans*, two Phoenix police officers stopped Mr. Evans for a routine traffic violation. Using a computer terminal in his police car, an officer ran a check for wants and warrants and discovered an outstanding misdemeanor arrest warrant for Mr. Evans. During the arrest, the officers found a bag of marijuana in Mr. Evans's car. The marijuana was seized as evidence.

It later turned out that, 17 days prior to the arrest, the court had quashed the Evans arrest warrant, but a clerk—at the court or with the police department—had failed to remove the warrant from the police computer. The trial court concluded that, because there had been no valid warrant for the arrest or the search that discovered the marijuana, Mr. Evans's Fourth Amendment protection against "unreasonable searches and seizures" was violated. The Arizona Supreme Court agreed.

The U.S. Supreme Court did not agree. The Court had argued, in *United States* v. *Leon,* that the exclusionary is intended to deter police misconduct, not to punish the errors of judges and magistrates. There was no deterrent benefit from excluding evidence seized pursuant to an improper warrant issued by a magistrate or one based on a faulty computer record. The limited good faith exception first issued in Leon "supports a categorical exception to the exclusionary rule for clerical errors of court employees."

The question left open is that of police department computer-generated information and records. Future cases are likely to address this area, as indicated by the comments of Justice O'Connor, who said the exclusionary rule would likely be "appropriate in a case in which the police utilized a computer recordkeeping system lacking a mechanism to ensure its accuracy. Any reliance on such a system would not be reasonable . . . Computers facilitate arrests, and police may have a constitutional duty to maintain the reliability of high-tech records."

Appendix
Technical Standards for Computing and Communications Security

With the aid of modern telecommunications technologies, vast quantities of data can be centrally maintained, processed, and organized and globally retrieved and put to use. Computer centers, as participants in powerful networks of personal computers and workstations, perform the role of gigantic file servers.

Such networks extend their global branches to all parts of the world. At the same time, they are rooted in a multitude of data-acquisition systems, each of which must channel vast quantities of data. Therefore, the tasks performed by the employees of a computer center encompass virtually all aspects of contemporary computer applications.

These tasks, however, cannot be accomplished without the incorporation of industry standards that create compatible formats and procedures. Accordingly, the planning and implementation concepts of modern computer centers are aimed toward the transparency and openness of the structures employed.

221

Standards Emerge

According to a report by International Resource Development, Inc., the emergence of standards is the greatest economic advance made by the computer industry in the last decade. It has accomplished what no amount of lawsuits against IBM could do to reduce the hold of proprietary system builders over the market. Standards such as ISDN, UNIX, OSI, and the IEE 802 family of LANs are increasingly important in terms of percentage of installations and value of new systems. The Internet largely uses formats of the Transmission Control Protocol/Internet Protocol (TCP/IP).

The major technical change in the standards field in the last decade is that de facto standards such as Bell modems have given way to formal standards, and standards committees no longer write up some company's existing product with a few variations as the standard for the rest of the industry.

This change to industry standards rather than proprietary products or vendor-initiated standards is the clearest measure of the extent of economic reorganization of the computer industry.

Standards-Making Organizations

The major standards organizations, as related to computers, data communications and security, are described next, along with a selection of standards relevant to computers and security.

American National Standards Institute (ANSI)

The ANSI is a voluntary, nongovernmental U.S. standards organization. Standards are produced through standards committees, technical committees, and task groups. The standards committee, X3, covers information systems; a technical committee, X3S3, deals with data communications. ANSI standards of interest in security include

- X3.92 data encryption algorithm;
- X3.105 data link encryption;
- X3.106 operation for the data encryption algorithm.

Copies of ANSI standards can be obtained from: American National Standards Institute, 1430 Broadway, New York, NY 10018.

Electronic Industries Association (EIA)

The EIA is a U.S. trade association, which produces standards on a broad range of electrical and electronic items. Technical Committee TR-30 covers data communications. Copies of standards can be obtained from Electronic Industries Association, Standards Sales, 2001 E St., NW, Washington, DC 20006.

International Telephone and Consultative Committee (CCITT)

The CCITT is a branch of the International Telecommunications Union, which is an agency of the United Nations. CCITT standards are issued as recommendations. These can be obtained from the U.S. Department of Commerce, NTIS, 5285 Port Royal Rd., Springfield, VA 22161.

International Organization for Standardization (ISO)

The ISO comprises the standards organizations of close to 90 countries. Some technical committees of the ISO cover computers, communications, and security. Copies of ISO standards can be obtained from ANSI at the address given previously.

Institute of Electrical and Electronic Engineers (IEEE)

The IEEE is a professional society of electronic and electrical engineers. The IEEE has standards committees on data communications and security. Copies of their standards may be obtained from IEEE Service Center, 45 Hoes Lane, Piscataway, NJ 08854.

Federal Information Processing Standards (FIPS)

The FIPS is a government standards-making organization under the Department of Commerce's National Institute of Standards and Technology (NIST). The FIPS often adopts ANSI standards for federal agency use. Or, it may adopt two families of standards, such as FIPS 161, Electronic Data Interchange (EDI), using X12, a U.S. standard, and EDIFACT, a set of international standards.

Copies of the FIPS standards are available for sale by the National Technical Information Service (NTIS), (703) 487-4650.

Data Encryption Standard

FIPS 46-2, the data encryption standard (DES), has been approved and the data encryption algorithm was reaffirmed in 1994 for five years.

Cryptographic Modules

FIPS-140-1, "Security Requirements for Cryptographic Modules," allows federal agencies to specify their security requirements for modules.

Secure Hash Standard

A "Secure Hash Standard," FIPS 180, is for use by federal agencies in protecting unclassified information. FIPS 180 specifies a secure hash algorithm (SHA), which can be used to generate a condensed representation of a message, or message digest. The SHA is required for use with the planned digital signature algorithm and whenever a secure hash algorithm is required for federal applications.

Key Management Standard

FIPS 171, "Key Management Using ANSI X9.17," adopts the ANSI standard for managing the keys used in secret-key encryption. FIPS 171 specifies a set of options for the automated distribution of keying material by the federal government using the ANSI X9.17 protocols.

Escrow Encryption Standard

FIPS 185, "Escrowed Encryption Standard" (EES), specifies a technology (the clipper chip) developed by the federal government and encryption/decryption keys to be escrowed.

According to the key escrow standard, it is designed to "assist law enforcement and other government agencies, under proper legal authority, in the collection and decryption of electronically transmitted information." Key escrowing was developed because "widespread use of encryption makes lawfully authorized electronic surveillance difficult."

The attorney general is charged with reviewing for legal sufficiency the procedures by which an agency establishes its authority to acquire the content of encrypted communications. "The privacy protections of the Constitution and relevant statutes afford adequate assurance of the efficacy of the standard."

The NIST estimates the cost of establishing the escrow system to be approximately $14 million. The cost of operating the key escrow facility is estimated to be $16 million annually.

The standard specifies the use of a symmetric-key encryption/decryption algorithm (SKIPJACK) and a law enforcement access field (LEAF) creation method (one part of a key escrow system), which provides for decryption of encrypted telecommunications. A key escrow system entrusts the two components making up a cryptographic key to two key component holders, called *escrow agents*.

The escrow agents provide the components of a key to a grantee (a law enforcement official) "only upon fulfillment of the condition that the grantee has properly demonstrated legal authorization to conduct electronic surveillance of telecommunications which are encrypted using the specific device whose device unique key is being requested."

Key escrow agents for the clipper chip are the Department of the Treasury's Automated Services Division and the Department of Commerce's National Institute of Standards and Technology.

The standard defines data as including "voice, facsimile and computer information communicated in a telephone system." A telephone system "is limited to a system which is circuit switched and operating at data rates of standard commercial modems over analog voice circuits or which uses basic-rate ISDN or a similar grade wireless service."

Digital Signature Standard

The "Digital Signature Standard" (DSS), FIPS 186, provides the capability to generate digital signatures that cannot be forged.

The standard specifies a digital signature algorithm (DSA), and the digital signature is a pair of large numbers represented in a computer as strings of binary digits. The digital signature is computed using a set of rules and a set of parameters such that the identity of the signatory and integrity of the data can be verified. The DSA provides the capability to generate and verify signatures.

Signature generation uses a private key to generate a digital signature. Signature verification uses a public key that corresponds to but is not the same as the private key. Each user possesses a private key and public key pair. Anyone can verify the signature of a user by employing that user's public key. Signature generation can be performed only by the possessor of the user's private key.

A hash function is used in the signature generation process to obtain a condensed version of data, called a *message digest*. The message digest is then entered into the DSA to generate the digital signature. The digital signature is sent to the intended verifier along with the signed message. The verifier of the message and signature verifies the signature by using the sender's public key. The same hash function also must be used in the verification process. The hash function is specified in a separate standard, FIPS 180. Similar procedures may be used to generate and verify signatures for stored as well as transmitted data.

The standard is applicable to all federal agencies and departments for the protection of unclassified information. The intent of the standard is to specify general security requirements for generating digital signatures. Conforming to the standard does not assure that a particular implementation is secure. Each agency or department is responsible for assuring that an overall implementation provides an acceptable level of security.

Security Label for Information Transfer

FIPS 188 defines a security label syntax for information exchanged over data networks and provides label encodings for use at the application and network layers. Security labels convey information used by protocol entities to determine how to handle data communicated between open systems. Security label information can be used to control access, specify protective measures, and determine additional handling restrictions required by a communications security policy.

Glossary

Access. The instruction, communication with, storage, or retrieval of data from a computer system or network.

Access control. The means of restricting entry or access to a computer room, terminal, or network.

ACH. Automated clearing house; facilities for the computer processing of large-volume financial transactions and the transfer, via electronic communication channels, of information of such transactions to member banks, bank customers, and others.

Alteration. Any material change of the terms of a writing fraudulently made by a party thereto.

Audit. A process of examining computer procedures to determine their reliability.

Authentic. Of undisputed origin or authorship; in being in accordance with, as stating fact; reliable; genuine; credible; authoritative and original (as in a document).

Authentication. The process of proving someone or something, such as a message or a transaction, is genuine or valid; in a computer system, verification of the identity of a user.

Authorization. A right, granted by management, to approve.

Beneficiary. In funds transfers, the person to be paid by the beneficiary's bank.

Bug. An error in a program or a system.

CAT. Credit authorization terminal; an electronic funds transfer (EFT) element usually located in retail outlets, which allows shoppers to get check cashing approval electronically.

Checksum. A short string of data that represents the content of an electronic document.

Computer network. A set of related, remotely connected devices and communication facilities including more than one computer system with the capability to transmit data among them through communicated facilities; the interconnection of communications lines (including microwave or other means of electronic communication) with a computer through remote terminals or a complex consisting of two or more interconnected computers.

Computer system. A set of related, connected or unconnected, computer equipment, devices, or computer software; a machine or collection of machines, used for governmental, educational, or commercial purposes, one or more of which contain computer programs and data, that performs functions including, but not limited to, logic, arithmetic, data storage and retrieval communication, and control.

Confidentiality. Ensuring that information is not made available or disclosed to unauthorized individuals. Electronic messages may be encrypted to assure confidentiality.

Consequential damages. The Uniform Commercial Code, Article 4A, limits the amount of damages businesses may recover from banks that commit errors. Unless there is an express agreement, banks cannot be held liable for consequent damages for errors made in connection with funds transfers.

Contract. An agreement between two or more parties that creates an obligation to do or not to do a particular thing. Three fundamental legal requirements of commercial contracts are an offer, acceptance, and consideration or payment.

Conversion. A change (in or to) another computer language, method, or equipment.

Credit transfer. A funds transfer where the instruction is given by the person making the payment; U.C.C., Article 4A governs these transfers.

Damages. Actual or compensatory damages are awarded for actual loss or injury, to compensate the victim. Punitive damages are awarded over and above a loss or injury and vary with the degree of punishment imposed.

Database. An organized collection of data processed and stored in a computer system.

Data coding. An identifier for a specific transaction; a code for a source document.

Data dictionary. Information that describes the structure and content of a database; the dictionary usually is stored in the computer.

Data manipulation. Changing data before or during the input process.

Debug. To detect, locate, and remove mistakes or malfunctions from a computer program or computer system.

Digital signature. An authentication method for computer messages.

Distributed database system. A computer system with the database distributed over several sites.

Documentation. All records on how a computer system was designed and how it operates.

Due care. The degree of care that a reasonable person would exercise to prevent the realization of harm, which under all the circumstances was reasonably foreseeable in the event that such care were not taken.

Electronic Funds Transfer Act (EFTA) of 1978 (Title XX, PL 95-630, 92 Stat. 3728, 15 U.S.C., Sect. 1693 et seq.). The federal statute that covers a wide range of electronic funds transfers, including point-of-sale transactions and other consumer payments. If any portion of a funds transfer is covered by EFTA, the whole funds transfer is excluded from U.C.C. 4A.

Evidence. Real evidence consists of tangible objects that are presented in court for the observation of the trier of fact as proof of the facts in dispute or in support of the theory of a party. Three other evidentiary forms are testimonial, documentary, and demonstrative.

Extortion. Coercion; obtaining property by inducing fear, using threat of or actual force or violence.

Fedwire. The Federal Reserve wire transfer network. Wire transfers made by Fedwire are governed by Federal Reserve Regulation J (12 C.F.R., Part 210).

File dump. A printout of the contents of a file.

File maintenance. An updating process for a file either to make changes or correct errors.

Foreseeability. The legal obligation of one party to protect another, second party against foreseeable, intentional wrongs done by a third party. "The probability of injury by one to the legally protected interests of another is the basis for the law's creation of a duty to avoid such injury, and foresight of harm lies at the foundation of the duty to use care and therefore of negligence. The broad test of negligence is what a reasonably prudent person would foresee and would do in the light of this foresight under the circumstances" (*American Jurisprudence*, 2d, Sect. 135).

Forgery. The fraudulent making or altering of an instrument that apparently creates or alters a legal liability on another.

Hard copy. Paper copy, such as a computer printout.

Hash total. A sum formed for control purposes by adding fields that normally are not related by a unit of measure. A verification method using a total compiled from numbers such as part numbers, invoice numbers, or customer account numbers.

Identification. A process that enables recognition of a user by a computer system; passwords and biometric systems are two methods of access control or identification.

Input. Data or instructions entered into the computer.

Integrity. An unbroken state; completeness; original perfect condition. The content of a message and whether it has been altered. The quality of being unimpaired, sound; fidelity and honesty. Message authentication techniques may be used to assure the integrity of electronic messages.

Intentionally. To do something purposely; desire to cause consequences.

Intermediary bank. A receiving bank other than the originator's bank or the beneficiary's bank.

Intimidation. Unlawful coercion; duress; putting in fear (of harm), the state must arise from the willful conduct of the accused.

Knowingly. To act with awareness of the nature of one's conduct; with knowledge; cognizant; having actual knowledge of or acting with deliberate ignorance of or reckless disregard for specific legal prohibitions.

Liability, directors and officers. Failure to perform a statutory or common law duty. Failure to use ordinary care and prudence, when it results in a loss, can generate liability. Special liabilities have been imposed by the Securities Acts of 1933 and 1934 and the Foreign Corrupt Practices Act.

Library. Physical location where data are stored; often referred to as the tape library.

Log. A record of what data processing has occurred; the log may be prepared manually or generated by computer.

Logic bomb. A computer program residing in a computer that is executed at appropriate or periodic times to determine conditions or states of a computer system and that facilitates the perpetration of an unauthorized act.

Memory. Storage, either in the central processing unit or in forms outside the computer.

Modem. Contraction of modulator-demodulator; converts data signals into voice signals and vice versa.

Negligence. A tort; the duty to use reasonable care—designed to protect persons from unintentional harm from the conduct of others. The criteria for determining negligence is vague.

Nonrepudiation. Unable to disclaim, renounce, or reject (as in an electronic message believed to be authentic).

On-line. A device or a process attached to or controlled by the computer.

Operating system. An integrated collection of computer programs resident in a computer that supervise and administer the use of computer resources to execute jobs automatically.

Parity. A method of verifying the integrity of transmitted data by checking the odd-even relationship of the bits.

Password. A protected word or string of characters that identifies a user, a specific resource, or an access type.

Payment order. An instruction by a business or the originator of the funds transfer to its bank to pay or credit an intended recipient a specific amount.

POS terminal. Point-of-sale terminal; used in retail outlets to log sales, it may be tied into a computer-based EFT system.

Privacy. A private matter, a secret; something kept or removed from public view or knowledge, belonging to or the property of a particular individual or group of persons.

Program. A set of instructions for the computer to perform a specific function or set of functions.

Prove. With regard to a fact, to meet the burden of establishing a fact as true by sufficient evidence; also "the burden of persuading the triers of fact that the existence of the fact is more probable than its nonexistence." Proof is the conclusion drawn from the evidence as to the existence of particular facts.

Prudent. Circumspect in action, or in determining any line of conduct (*Black's Law Dictionary*).

Questioned document. A document that has been disputed in whole or in part in respect to its authenticity, identity or origin.

Reasonable. Fit and appropriate to the end in view (*Black's Law Dictionary*).

Reckless. Careless, heedlessly indifferent to consequences; lacking in due caution.

Recoverability. A control objective of contingency plans, data or information backup and retention to minimize downtime and ensure operational recovery of information and communications systems.

Reliability. Meeting management control objectives of accuracy, completeness, timeliness, security, auditability, and recoverability. Information systems may have to meet a definition of reliability based on a set of control objectives.

Run. To execute a computer program.

Schedule. List of actions to be executed for a set of computer transactions.

Scienter. Knowledge, referring to those wrongs or crimes that require a knowledge of wrong in order to constitute the offense.

Sender. Includes the customer in whose name a payment order is issued if the order is the authorized order of the customer under Universal Commercial Code (UCC) 4A-203 subsection (a), or it is effective as the order of the customer under UCC 4A-203 subsection (b).

Signature. The traditional way of making a document, such as a contract, legally effective. A signing may be a symbol with present intention to authenticate a writing.

Simulation and modeling. The creation of a system to duplicate one already in existence, a parallel system used to simulate information, reports, and data.

Source document. A form for recording data, usually the first record.

Spoliation. Of Roman origin, from *spoliarium,* a room in the amphitheater where slain gladiators were stripped of their armor, weapons, and clothes. To spoliate is to spoil or despoil, an act of plundering; in modern law, the act of destroying a document or of injuring or tampering with it as to destroy its value as evidence. Spoliation is punished as an obstruction of justice offense.

System flowchart. A diagram showing the flow of data.

Tampering. The unauthorized modification of a computer device or system that causes its degradation or malfunction.

Terminal. A wide variety of telephones, consoles, PBXs, data transmission, or other communication devices used to terminate one or more telephone circuits or data transmission cables.

Timeliness. A control objective of avoiding disruption or delays in data processing or transmission.

Tort. A private injury or wrong arising from a breach of duty created by law.

Transaction log. A printout delineating all interactive input, process and outputs on any file.

Trojan horse. Computer instructions secretly inserted in a computer program so that, when it is executed in a computer, unauthorized acts are performed.

Update. To bring a computer file up to date.

Verification. A check for accuracy; confirmation of correctness and authenticity by sworn or equivalent confirmation or truth.

Virus. A self-propagating Trojan horse.

Wholesale wire transfers. Transfer payments principally between businesses or financial institutions.

Wiretapping. Interception of data communications signals with the intent to gain access to data transmitted over communications circuits.

Worm. A form of malicious, self-replicating code.

Wrongful acts. " . . . any act which in the ordinary course (of business) will infringe upon the rights of another to his damage . . . " (*Black's Law Dictionary*).

Bibliography
Selected Publications on Protection, Law, and Computing

American Jurisprudence, 2nd ed. Rochester, NY: Lawyers Cooperative Publishing Co.

American Law Institute. *Restatement Second of Torts*. St. Paul, MN: American Law Institute Publishing, 1965.

Androphy, J. *White Collar Crime*. Colorado Springs, CO: Shepard's/McGraw-Hill, 1992.

Arkin, Stanley, et al. *Business Crime: Criminal Liability of the Business Community*. New York: Matthew Bender, 1981.

Arkin, Stanley, et al. *Prevention and Prosecution of Computer and High Technology Crime*. Albany, NY: Matthew Bender, 1989.

Bailey, F. Lee, and Henry B. Rothblatt. *Investigation and Preparation of Criminal Cases*. Rochester, NY: Lawyers Co-operative, 1970.

Baker, D., and R. Brandel. *Law of Electronic Funds Transfer Systems*. New York: Warren, Gorham & Lamont, 1988.

Bender, David. *Computer Law: Evidence and Procedure*. New York: Matthew Bender, 1978.

Bernacchi, Richard L. *A Guide to the Legal and Management Aspects of Computer Technology*. Boston: Little, Brown and Co., 1986.

Block, Dennis J., and Marvin J. Pickholz. *The Internal Corporate Investigation*. New York: Practicing Law Institute, 1980.

Bologna, Jack. *Computer Crime: Wave of the Future.* Madison, WI: Assets Protection, 1981.

Bologna, Jack. *Handbook on Corporate Fraud.* Stoneham, MA: Butterworth, 1993.

Bologna, Jack, and Paul Shaw. *Fraud Awareness Manual.* Madison, WI: Assets Protection, 1995.

Brickey, Kathleen. *Corporate Criminal Liability.* Deerfield, IL: Callaghan & Co., 1989.

Brinson, Dianne J., and Mark F. Radcliffe. *Multimedia Law and Business Handbook.* Menlo Park, CA: Ladera Press, 1996.

Brown, L. *The Legal Audit: Corporate Internal Investigation.* New York: Clark Boardman Co., 1990.

Business Conference Board. *Corporate Ethics.* New York: The Conference Board, 1990.

Comer, Michael. *Corporate Fraud.* New York: McGraw-Hill, 1977.

Committee of Sponsoring Organizations of the Treadway Commission (COSO). *Internal Control—Integrated Framework.* New York: The Committee of Sponsoring Organizations of the Treadway Commission, 1992.

COSO. *Reporting to External Parties* (addendum). New York: COSO, 1994.

Coughran, Edward. *Computer Abuse and Criminal Law.* San Diego: Computer Center, University of California, 1976.

Dorr, Robert C., and Christopher H. Munch. *Protecting Trade Secrets, Patents, Copyrights, and Trademarks,* 2nd ed. New York: John Wiley & Sons, Inc., 1995.

Fischer, L. Richard. *The Law of Financial Privacy,* 2nd ed. New York: Warren, Gorham & Lamont, 1991.

Forensic Services Directory. Princeton, NJ: National Forensic Center (annual).

Frank, P., ed. *Litigation Services Handbook.* New York: John Wiley & Sons, 1990.

Fricano, E., ed. *Corporate Practice Series Guide to RICO.* Washington, DC: Bureau of National Affairs, 1986.

Glekel, Jeffrey, ed. *Business Crimes: A Guide for Corporate and Defense Counsel.* New York: Practicing Law Institute, 1982.

Goldblatt, M. *Preventive Law in Corporate Practice.* New York: Matthew Bender, 1991.

Hannon, L. *Legal Side of Private Security.* Westport, CT: Greenwood Publishing Group, 1992.

Hartsfield, H. *Investigating Employee Conduct.* Deerfield, IL: Callaghan & Co., 1988.

Keeton, W. Page, et al. *Prosser and Keeton on the Law of Torts,* 5th ed. St. Paul, MN: West Publishing Co., 1984.

Kell, William G., and Robert K. Mautz. *Internal Controls in U.S. Corporations.* New York: Financial Executives Institute Research Foundation, 1980.

Kramer, M. W. *Investigative Techniques in Complex Financial Crimes.* Washington, DC: National Institute on Economic Crime, 1989.

Krauss, Leonard I., and Aileen MacGahan. *Computer Fraud and Countermeasures.* Englewood Cliffs, NJ: Prentice-Hall, 1979.

Kropatkin, Philip, and Richard P. Kusserow. *Management Principles for Assets Protection: Understanding the Criminal Equation.* New York: John Wiley & Sons, 1986.

Loeffler, Robert M. *Report of the Trustee of Equity Funding Corporation of America.* U.S. District Court for the Central District of California, October 31, 1974.

Longley, D., M. Shain, and W. Caelli. *Information Security Dictionary of Concepts, Standards and Terms.* New York: Stockto Press.

Milgrim, Roger M. *Milgrim on Trade Secrets.* Albany, NY: Matthew Bender, 1994.

Miller, Gordon H. *Prosecutor's Manual on Computer Crimes.* Decatur, GA: Prosecuting Attorneys' Council on Georgia, 1978.

National Commission on Fraudulent Financial Reporting. *Exposure Draft.* Washington: National Commission on Fraudulent Financial Reporting, 1987.

Nimmer, Raymond T. *The Law of Computer Technology,* rev. ed. New York: Warren, Gorham & Lamont, 1994.

Obermaier, H. *White Collar Crime.* New York: Law & Seminars Press, 1990.

O'Neill, Robert. *Investigative Planning.* Report prepared for Battelle Law and Justice Study Center, Seattle, WA, 1978.

Parker, Donn B. *Crime by Computer.* New York: Charles Scribner's Sons, 1976.

Perritt, Henry H., Jr. *Law and the Information Superhighway.* New York: John Wiley & Sons, Inc., 1996.

Renesse, Rudolf Van. *Optical Document Security.* New York: John Wiley & Sons, 1995.

Roddy, K. *RICO in Business and Commercial Litigation.* Colorado Springs, CO: Shepard's/McGraw-Hill, 1992.

Schabeck, Tim. *Computer Crime Investigation Manual.* Madison, WI: Assets Protection, 1979.

Seidler, Lee J., Frederick Andrews, and Marc J. Epstein. *The Equity Funding Papers: The Anatomy of a Fraud.* Santa Barbara, CA: John Wiley & Sons, 1977.

Sigler, J. and J. Murphy. *Corporate Lawbreaking and Interactive Compliance.* Westport, CT: Greenwood Publishing Group, 1991.

Skupsky, D. *Legal Requirements for Microfilm, Computer, and Optical Disk Records.* Denver: Information Requirements Clearinghouse, 1994.

Skupsky, D. *Recordkeeping Requirements.* Denver: Information Requirements Clearinghouse, 1989.

Soble, Ronald L., and Robert E. Dallos. *The Impossible Dream: The Equity Funding Story, the Fraud of the Century.* New York: Putnam, 1975.

Soma, John T. *Computer Technology and the Law.* Colorado Springs, CO: Shepard's/McGraw-Hill, 1983 and 1995 Supplement.

U.S. Sentencing Commission. *Federal Sentencing Guidelines Manual.* Washington, DC, 1994.

Vaughan, D. *Controlling Unlawful Organizational Behavior—Social Structure and Corporate Misconduct.* Chicago: University of Chicago Press, 1983.

Vergani, James V., and Virginia V. Shue. *Fundamentals of Computer-High Technology Law.* Philadelphia: American Law Institute/American Bar Association, 1991.

Villa, J. *Banking Crimes: Fraud, Money Laundering, and Embezzlement.* New York: Clark Boardman, 1988.

Wagner, Charles. *The CPA and Computer Fraud.* Lexington, MA: Lexington Books, 1979.

Wasik, M. *Crime and the Computer.* Cary, NC: Oxford University Press, 1991.

Wood, Charles W. *Information Security Policies Made Easy.* Sausalito, CA: Baseline Software, 1994.

Wright, Benjamin. *The Law of Electronic Commerce: EDI, E-Mail, and Internet,* 2nd ed. Boston: Little, Brown & Company, 1995.

Wright, Benjamin. "The Legality of the PenOp Signature," unpublished paper, 1995.

Index

abuse of trust, 178–179
access control, 104
advertising
 false, 151
 unsolicited fax, 204–205
Age Discrimination in Employment Act, 82
Agreement on Trade-Related Aspects of
 Intellectual Property Rights (TRIPS
 Agreement), 35–37
American National Standards Institute
 (ANSI), 222
Americans With Disabilities Act (ADA)
 record keeping for, 98, 101
Americans with Disabilities Act of 1990
 (ADA)
 definition of disability, 131–132
 disability-related terminology, 132
 preemployment inquiries and, 133
Americans With Disabilities Act of 1992, 87
Annunzio-Wylie Anti-Money Laundering
 Act of 1992, 95
Anticounterfeiting Consumer Protection
 Act of 1996, 56–57
antitheft policies, 30–32
Arizona v. *Evans*, 220
Armstrong v. *Executive Office of the
 President*, 98
ATM safety, 4
audits
 compliance, 25
 detecting fraud through, 29
 electronic commerce, 122–123

of e-mail policies, 141
of software copying, 75–76
of trade secret programs, 40
Automated Clearing House (ACH), 96

Bank Secrecy Act of 1970, 95
banking. *See also* electronic commerce
 funds transfers, 95
 and record keeping, 94, 95–96
Berne Convention for the Protection of
 Literary and Artistic Works, 36
brand equity, 51–52
bulletin boards (BBS), on-line
 copyright/trademark infringements,
 69–70
 search warrants for, 216–217

Cable Act of 1992, 82–83
Cable Communications Policy Act of
 1984, 82–83
CALEA (Communications Assistance for
 Law Enforcement Act), 83
Caremark International Inc., 23
CCITT (International Telephone and
 Consultative Committee), 223
Chaplinsky v. *New Hampshire*, 130–131
China, copyright abuse in, 35
civil investigative demands (CIDs), 218
Civil Rights Act of 1964, 87, 101
collateral prosecutions, 10–11
collective/aggregate knowledge, 9
color, as trademark, 53–54

239